Complete Guide
to Fertility

Marion Powell
Women's Health Information Centre

2897

 The American Society for Reproductive Medicine

Complete Guide
to *Fertility*

SANDRA ANN CARSON, M.D., AND PETER R. CASSON, M.D.,

WITH DEBORAH J. SHUMAN

Produced by The Philip Lief Group, Inc.

CB
CONTEMPORARY BOOKS

Library of Congress Cataloging-in-Publication Data

Carson, Sandra A. (Sandra Ann)
     The American Society for Reproductive Medicine complete
guide to fertility / Sandra Ann Carson and Peter R. Casson, with
Deborah Shuman.
          p.   cm.
     Includes index.
     ISBN 0-8092-2862-9
     1. Infertility—Popular works.   I. Casson, Peter R.   II.
Shuman, Deborah.   III. American Society for Reproductive
Medicine.   IV. Title   V. Title: Complete guide to fertility.
     RC889.C365   1999
     618.1'78—dc21   98-50746
                                                                              CIP

Cover design by Monica Baziuk
Interior design by City Desktop Productions, Inc.

Published by Contemporary Books
A division of NTC/Contemporary Publishing Group, Inc.
4255 West Touhy Avenue, Lincolnwood (Chicago), Illinois 60646-1975 U.S.A.
Printed in the United States of America
International Standard Book Number: 0-8092-2862-9
99 00 01 02 03 04 QP 15 14 13 12 11 10 9 8 7 6 5 4 3 2 1

ASRM Executive Director

J. Benjamin Younger, M.D.

ASRM Medical Director

Roger D. Kempers, M.D.

Contributing Editors

Larry I. Lipshultz, M.D., Department of Urology
Baylor College of Medicine, Houston, Texas
ASRM President, 1999

John A. Rock, M.D., Department of Obstetrics and Gynecology
Emory University School of Medicine, Atlanta, Georgia
ASRM Past President

ASRM Patient Education Committee

Anita Maria Aloia, L.V.N.
Santa Monica, California

Sharon N. Covington, M.S.W.
Rockville, Maryland

Carole L. DeMaio, R.N.C.
Winter Park, Florida

Timothy N. Hickman, M.D.
Wilford Hall Medical Center, San Antonio, Texas

Bradley S. Hurst, M.D., Department of Obstetrics and Gynecology
University of Colorado Health Science Center, Denver, Colorado

Jonathan P. Jarow, M.D., Department of Urology
Johns Hopkins Hospital, Baltimore, Maryland

Paul B. Marshburn, M.D., Department of Obstetrics and Gynecology
Carolina Medical Center, Charlotte, North Carolina

Stephen G. Somkuti, M.D., Ph.D.
Abington, Pennsylvania

Samuel Smith, M.D., Department of Obstetrics and Gynecology
Sinai Hospital of Baltimore, Baltimore, Maryland

Contents

Acknowledgments

The editors wish to express their deep appreciation to Suzi Lindsay for her skills, earnest attention to detail, and constant oversight in preperation of this book.

Complete Guide to Fertility

1

An Introduction to Infertility

Through most of human history, the reproductive cycle has been shrouded in mystery. At various times and in various cultures, pregnancy was thought to be set in motion by the gods, by dreams, by swimming in certain waters, by dancing. Even when people accepted the connection between sex and pregnancy, fertility was still a mysterious concept. At the turn of the twentieth century, couples still depended on rumors, old wives' tales, home remedies, herbs, incantations, and elixirs to aid fertility and solve their infertility.

At the turn of the twenty-first century, the situation is very different. Breakthroughs in the science of fertility are regularly featured in major newspapers, magazines, and television news programs. The problem is no longer a lack of knowledge, but rather an abundance of information that can be overwhelming and confusing.

If you have picked up this book, you may be among the millions of Americans who are trying to understand the latest in infertility research.

You know that hope lies in medical science, but navigating the sea of information to find the scientific facts *you* need seems like an overwhelming task.

Here is a landmark book to guide you. The American Society for Reproductive Medicine (ASRM) is the central repository for information on infertility. The physicians responsible for *The American Society for Reproductive Medicine Complete Guide to Fertility* are board-certified reproductive endocrinologists who have analyzed the most up-to-date research results, have separated facts from wishful thinking, and have placed the facts in this one volume to provide the information you need in a form you can understand.

If you have been trying without success to become pregnant, you are almost certainly awash in emotions ranging from confusion to anger and fear. This book is designed to help you take rational steps toward fulfilling your dream of having a baby, and to help you sort out your emotions and care for yourself in the process.

What Is Infertility?

As we'll see in Chapter 2, "The Basics of the Reproductive Process," timing is everything when it comes to conception. But even when timing is perfect, it is not always possible to predict when pregnancy will occur. Some couples get pregnant the first time they have unprotected sex. Others conceive only after years of trying. Given that this is the case, at what point can you tell whether you and your partner have an infertility problem?

Definition of Infertility

The accepted definition of *infertility* is the absence of conception after at least one year of regular intercourse without birth control. ("Regular" intercourse means two or three times every week.) About 7 percent of all married couples, including more than 10 million women, met this definition of infertility in 1995. Of all couples trying to get pregnant, about 15 percent have not succeeded within one year.

How Common Is Infertility?

The news coverage of stories related to infertility would make it seem as though the rate of infertility is rising in the United States and other developed nations. There is a general notion that, especially because more women are delaying childbirth until their 30s or even 40s, there are more cases of infertility today than ever before.

Is there an infertility epidemic? It is true that more women are seeking infertility services—about 15 percent of all women of reproductive age (15 to 44). But many experts think that more women are seeking these services simply because there are more infertility specialists available today.

In fact, statistics show that the relative number of couples seeking help with infertility is about the same today as it was 30 years ago. The number of childless women in their early 20s has actually declined in the past 20 years. Infertility is still more common in women over age 30, but birth rates in all age groups over 30 have risen since the mid-1970s.

Still, if you are having trouble conceiving, you are clearly not alone. As you explore the possible causes of and treatments for infertility, it may help you to become aware of the many resources available to you. In this chapter you will find a general overview of the types of infertility and its causes, as well as information on the expanding field of reproductive medicine and the screening process for infertility.

Types of Infertility

To state the obvious, there is only one way for a child to be conceived: a sperm must fertilize an egg. However, there are many ways in which this process can be hindered or prevented. The types of infertility described here are separated into male and female factors. Both men and women can have congenital anomalies—birth defects that cause infertility. Both partners can be infertile because of surgery, disease, or an infection that has damaged some part of the reproductive system.

Male factors cause 30 to 40 percent of all cases of infertility; female factors cause roughly 40 percent of cases. About 15 percent of cases are due to a combination of male and female factors. In 5 to 10 percent of cases, no cause for the infertility can be found.

Experts describe infertility as either *primary* or *secondary*. A couple who has never achieved pregnancy is experiencing *primary infertility*, whereas a couple who has had a pregnancy in the past (whether or not the pregnancy resulted in a living child) is experiencing *secondary infertility*. For any couple facing the inability to conceive, however, this is most likely a meaningless distinction. Whether they already have children or not, the most important fact remains the same: They want to get pregnant, but so far have not been able to do so.

Male Infertility

To fertilize an egg, sperm produced in the testes must be present in large numbers in the semen. The sperm must also be motile—able to move far enough and fast enough to navigate the female reproductive system—and strong enough to break through the outer layers of an egg. In some men, sperm are not present in the semen or are present only in small quantities. This condition may be the result of injury, surgery, or a congenital defect.

Any number of influences can cause congenital defects. Among the most common are:

- Prenatal exposure to DES (*diethylstilbestrol*), a medication that was given to some pregnant women in the 1950s to prevent miscarriage

- Exposure to radiation, certain pesticides, or heavy metals (such as lead or mercury)

- Diseases that reduce the body's ability to produce sperm (men who contract mumps after they have reached puberty, for example, are at risk for sterility)

To determine whether a couple's infertility is due to a male factor, a specialist will perform a physical exam to detect any abnormalities of the man's reproductive system. Tests may also be done to evaluate the quantity, motility, and function of his sperm.

Female Infertility

Conditions that contribute to or cause infertility in women are usually either anatomical abnormalities or malfunctions in the reproductive system. Chapter 2 describes the female reproductive system in detail, and Chapter 4 discusses the various causes of female infertility. Here, however, is an overview of complications a woman might experience, which can be broadly classified as tubal, peritoneal, ovulatory, uterine, and cervical factors.

Tubal Factors Irregularities of the fallopian tubes account for an estimated 35 percent of female infertility. The fallopian tubes are the thin, tubelike structures that lead from the ovaries to the uterus. A fertilized egg travels from the ovary through these tubes on its way to the inside of the uterus. Certain diseases and infections, most notably *pelvic inflammatory disease* (PID), can leave scar tissue that partially or completely blocks the fallopian tubes.

Tests can be performed to determine whether the fallopian tubes are open and healthy. During one of these tests, *hysterosalpingography* (HSG), the doctor introduces a special dye through the cervix into the uterus and tubes. He or she can then view the size and shape of these structures on a monitor somewhat like an x-ray machine. Sometimes surgery can correct tubal problems, but most often women with this type of infertility must consider alternative methods to achieve pregnancy. (See Chapters 4 and 5 for diagnosis and treatment of tubal infertility.)

Peritoneal Factors In about another 35 percent of cases of female infertility, the cause can be traced to a disease or disorder of the *peritoneum* (abdominal cavity). *Endometriosis* is one of the more common of these disorders. With endometriosis, tissue that resembles the lining of the uterus is found elsewhere in the abdomen. For reasons not wholly understood, women with endometriosis are about twice as likely to be infertile as those without this condition.

Another peritoneal factor that causes infertility is scar tissue, or *adhesions*. Scar tissue sometimes forms after abdominal or pelvic surgery or infection. This tissue can adhere to and bind together some of the structures inside the pelvis, interfering with their normal function.

Peritoneal factors can be diagnosed with *laparoscopy*, a minor surgical procedure that allows the doctor to view the inside of a woman's abdomen. In many cases of endometriosis, doctors prescribe hormones to help shrink the abnormal tissue. If necessary, surgery can be performed to remove adhesions.

Ovulatory Factors In most women, one egg is released by one of the ovaries each month about halfway through the menstrual cycle. Problems with ovulation are the cause of about one-quarter of cases of female infertility. Since hormones trigger the release of eggs from the ovaries, irregular ovulation—or a lack of ovulation altogether—usually can be linked to hormone imbalances.

Tests can determine whether a woman is producing reproductive hormones in sufficient quantities at the right time to stimulate the release of an egg each month. If not, specialists often prescribe drugs to trigger or enhance ovulation. Chapter 5 presents details on treating ovulatory problems.

Uterine Factors Abnormalities of the uterine cavity cause about 5 percent of all cases of female infertility. Most often, these problems are due to benign (noncancerous) growths inside the uterus. *Uterine fibroids* and *polyps* are growths that may be attached to the outer or inner wall of the uterus or may form inside the uterine wall itself. Some fibroids and polyps can be treated with hormones; others can be corrected with surgery.

Cervical Factors To reach and fertilize a woman's egg, sperm must be able to pass through the mucus produced by the cervix (the lower opening of the uterus). In some women, the texture and consistency of the cervical mucus do not allow the passage of sperm in great enough quantity to allow fertilization. This problem may be caused by prior surgery of the cervix. Cervical mucus can also be toxic to sperm. Researchers have discovered high toxicity levels, for example, in women who smoke.

Cervical factors are rarely the sole cause of female infertility. Most often these problems occur along with some other abnormality, such as fibroids or endometriosis. When a cervical factor contributes to infertility, it may be possible to treat it with antibiotics or hormones, or to bypass the mucus by performing artificial insemination (see Chapter 5).

Unexplained Infertility

In about 5 to 10 percent of couples who seek treatment for infertility, no cause is ever discovered. This can be an especially difficult time for a couple who has undergone a number of diagnostic tests, only to find that all the results have come back normal and the problem remains. Many couples choose to continue with more complex testing and to try the many treatment options available to them. Chapter 5 goes into detail regarding such tests and options for couples who can find no obvious cause for infertility.

Infertility Medicine

Many couples faced with infertility seek the services of doctors who specialize in diagnosing and treating the various causes. Many of these doctors have become certified to practice medicine in *reproductive endocrinology*. Reproductive endocrinologists have years of highly specialized, advanced training and experience in the diagnosis and treatment of problems in both male and female reproduction. They have taken instruction in all aspects of infertility medicine, including complex infertility surgeries, medication to treat ovulatory problems, and treatment of infertility with techniques of assisted reproduction.

The field of reproductive endocrinology is a branch of the discipline of obstetrics and gynecology (ob/gyn). In addition to four years of medical school, ob/gyns have completed another four years of specialized training. To become certified to practice reproductive endocrinology, an ob/gyn must further complete a two- to three-year fellowship in this subspecialty and pass further exams.

How do you know when you and your partner should seek counseling, and possibly treatment, for infertility? As mentioned previously, the standard definition of infertility is the failure to achieve pregnancy after one year of regular, unprotected intercourse. Because there is a natural decline in a woman's fertility as she reaches her late 30s and early 40s, physicians usually advise women over the age of 35 to begin seeking advice and treatment after six months of trying to get pregnant. If you and your partner have been trying to conceive for at least one year—or six months if you are over 35—the information in this book will give you the background you need to choose an infertility specialist and will help ensure that you get the treatment you need.

Screening

What can one expect during the initial visit to an infertility specialist? First, the physician will gather a detailed medical history from both partners. He or she will then discuss the reproductive process in detail and assess whether some of the couple's lifestyle practices may be causing or contributing to infertility. A physical exam is also done on both partners. The goal of the first visit is to determine whether there may be a simple explanation that can be addressed without costly treatments. Some possible explanations that the doctor might suggest are discussed below.

Timing of Intercourse Although getting pregnant seems simple, it actually requires that a complex sequence of events occur at just the right time and in just the right order. Even for women in their prime reproductive years, the chances of pregnancy occurring during each monthly cycle is only about 20 to 25 percent.

The average woman ovulates about halfway through a 28-day cycle. A woman who knows that her cycle is regular can maximize her chances of conceiving by timing intercourse to coincide with ovulation. Even when the cycle is not as regular as 28 days, it is still possible for a couple to know when ovulation occurs. During the initial screening with a reproductive specialist, if the physician perceives that the couple has not learned to time intercourse with the woman's fertile periods, he or she will counsel both partners on how to identify the signs that indicate ovulation. Often, this process of education is all it takes to solve a couple's infertility.

Frequency of Intercourse To optimize their chances of becoming pregnant, a couple should have intercourse at least two to three times a week. When intercourse occurs less often than this, the odds of pregnancy may be lowered. Increasing the frequency of intercourse may solve the problem.

If a couple is not having regular intercourse, the reasons need to be explored. The partners may simply be too busy or too tired for sex. There may be problems in their relationship that need to be resolved before bringing a child into their lives. It is not uncommon for people to think that marital problems will be diminished with the birth of a baby. But statistics show that more often, the demands of caring for a new life put added pressures on an already stressed relationship.

Other Sexual Factors Painful intercourse, or *dyspareunia*, may contribute to infertility. If sex is uncomfortable or painful for the woman, a couple may not have sex often enough to maximize their chances of becoming pregnant. When intercourse does take place, it may not last long enough for the man to ejaculate inside his partner because of the woman's discomfort.

If painful intercourse is an issue, the doctor will evaluate the problem to determine if it has a medical cause that can be treated. In rare instances, emotional problems between a couple or in the woman herself may contribute to painful intercourse. Counseling or psychotherapy for both partners will be indicated in such cases.

Lifestyle Factors During the screening process, a fertility expert will ask a couple about their use of alcohol, tobacco, and other drugs and medications. It is important that the couple report honestly about the use of these substances, since they can inhibit fertility. The doctor will also assess the woman's diet and weight. If the woman is extremely overweight or underweight, the doctor might suggest that these factors are contributing to infertility.

Smoking decreases fertility in both men and women. A woman who smokes is three to four times more likely than a nonsmoker to have a one-year or greater delay in conceiving. The more cigarettes she smokes, the lower her chances of becoming pregnant. When a couple decides to become pregnant, both partners should stop smoking— ideally, before they stop using birth control.

The link between delayed conception and *caffeine, alcohol, and other drugs*, including illicit drugs such as *cocaine, tranquilizers, and amphetamines ("speed")*, is unclear. However, the medical problems that result from addiction to these substances are known to have adverse effects on a woman's ability to conceive. Some are proven to be harmful to a fetus, especially within the first weeks after conception. As part of optimizing their health, couples should stop or refrain from using alcohol, cigarettes, and other drugs before trying to conceive.

Because *obesity* can impair ovulation, women wishing to become pregnant should maintain their weight between 80 and 130 percent of their ideal body weight (see Chapter 2, page 29, to determine ideal body weight). For some obese women, losing weight alone can lead to pregnancy.

Low weight or *rapid weight loss* can lead to a decrease in an important hormonal message that the brain sends to the ovaries in women and the testes in men. If low weight or rapid weight loss is identified as the cause of infertility, the preferred treatment is to stop losing weight, or even to gain weight if needed.

Exploring Your Options

If none of the factors previously discussed is relevant to you, or if you have tried to address some of the simpler explanations for infertility with no results, then you are probably ready for a more in-depth medical evaluation to find an effective treatment. Before embarking on this course, however, both you and your partner should be aware of what to expect in terms of medical evaluations, the time and commitment that will be needed, and the cost of diagnosis and treatment. The information presented here, coupled with thoughtful discussions by infertility specialists, can help you sort out the many issues and factors you'll need to consider.

In the next chapter, you'll find detailed information about the male and female reproductive systems, the process of reproduction, and the factors that can interfere with or prevent this process. You may already know a great deal about reproduction. However, new research has revealed many aspects of reproduction never before understood. You and your partner will enhance your chances of success by educating yourselves with the latest information and making use of all the knowledge, technology, and techniques available to you.

2

The Basics of the Reproductive Process

The complex process of human reproduction requires that multiple events occur in a specific order and with precise timing. A change or interruption at any step along this complicated pathway can impede or prevent conception.

This chapter reviews the anatomy and function of the human reproductive system and the process of conception. Also explained are some of the variables that can hinder fertility. You will see that some factors, such as medical conditions, are not within your control and require physician-directed medical solutions. But other factors represent lifestyle choices that you can change if they apply to you.

The Female Reproductive System

The primary purpose of a woman's reproductive system is to conceive and maintain the life of a fetus. Each part of the system plays a specific role in the fetus's creation and nourishment.

Basic Anatomy

The female reproductive system is comprised of the following organs and structures (see Figure 2.1):

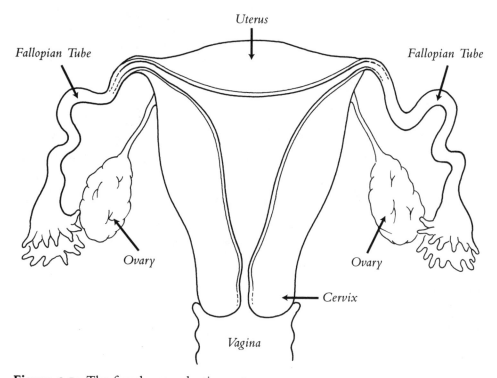

Figure 2.1: The female reproductive system.

- The *ovaries* are two small organs on either side of the lower pelvis. They contain the *ova*, or eggs, that can become fertilized and grow into a fetus. A mature ovum is just barely visible to the naked eye if it is shown against a bright background.

- The *fallopian tubes* are narrow tubes that lead from each ovary to the uterus. Fertilization of an egg normally occurs in one of the fallopian tubes as the egg travels toward the uterus; the tube also provides an "incubator" for the embryo before implantation.

- The *uterus* is the muscular organ in which a developing fetus grows and is nourished during pregnancy. During conception, an embryo attaches to the *endometrium*, or lining of the uterus. The uterus is extremely elastic: Before pregnancy, it weighs only about two ounces and can hold a volume of less than half an ounce. Near the end of pregnancy, it weighs about two-and-a-half pounds and has stretched to hold the fully grown fetus and the additional tissue and fluid needed to sustain the pregnancy.

- The *cervix* is the lower, narrow "neck" of the uterus that opens into the vagina. During labor, the cervix expands to allow the fetus to pass through. A small amount of mucus is normally present in the cervical opening. During pregnancy, this fluid congeals to form a plug that helps protect the opening to the uterus against infection.

- The *vagina* is the muscular passageway that leads from the cervix to the outside of the body. A baby passes through the cervix and vagina (sometimes called the *birth canal*) during a vaginal birth. The vagina is lined with mucous membrane tissue that stretches and elongates during childbirth.

- The *breasts* contain glands that produce milk to nourish the infant after birth. During pregnancy, hormones produced by the placenta signal breast tissue to begin producing milk. Stimulation of the breasts by the baby's suckling ensures the continued release of these hormones and production of milk throughout the nursing period.

The Menstrual Cycle

The entire reproductive process is directed and coordinated by hormones secreted by the *hypothalamus* and *pituitary*. These hormones control the menstrual cycle, which begins each month with the shedding of blood and tissue from the uterus. By this mechanism nature provides continual opportunities for reproduction to occur throughout a woman's prime childbearing years.

A Word on Hormones Hormones are chemical messengers that coordinate the function of all of the body's organ systems. They are produced by the *endocrine glands*, specialized organs that release and respond to hormones in the blood. The hypothalamus and pituitary—located at the base of the brain—are the endocrine glands that produce hormones that stimulate the ovaries. The ovaries, in addition to producing eggs, secrete sex hormones— *estrogen* and *progesterone*. Other hormone-producing glands are the *thyroid*, situated at the base of the throat, and the two *adrenal glands*, which lie next to the kidneys.

How Ova Are Formed A woman's ovaries are formed while she is in the fetal stage of her life, during the first trimester of her mother's pregnancy. Cells that will become ova are present in the fetus at about the sixth week of pregnancy. Six months after conception, the fetal ovaries already contain millions of these future ova.

The cells of all living things contain *chromosomes*, submicroscopic structures carrying the genes that determine physical traits such as height and eye color. Most cells of the human body contain forty-six chromosomes. The exceptions are the *germ cells*, or sex cells—the ova (in females) and the sperm (in males). The sex cells are distinct in that they contain only half as many (twenty-three) chromosomes. At fertilization, the sperm and the ovum combine their chromosomes to make up the full forty-six chromosomes of the fetus.

Germ cells continue to develop throughout the fetus's life. As these cells multiply, some are destined to become eggs, whereas others will eventually dissolve and be absorbed back into the body. The number of germ cells thus continues to decrease over time, both before and after birth. By the time she is born, a baby girl has two to four million germ cells in her ovaries. At puberty, less than half a million remain. If a female never becomes pregnant and has regular periods until she reaches menopause, a total of about 300 to 400 ova are released by her ovaries throughout her life. The rest are reabsorbed.

Ovulation Inside each ovary, the eggs themselves are contained in structures called *follicles*. Each month, about twenty follicles are "primed" for ovulation as the hypothalamus sends out *gonadotropin-releasing hormone* (GnRH) to the pituitary. In response, the pituitary releases *follicle-stimulating hormone* (FSH). This hormone in turn triggers one, or occasionally more, of the follicles to grow and develop in preparation for ovulation. (It is not completely understood how one of the follicles is "chosen" for ovulation.) At this point, the immature (not-yet-released) ovum inside the follicle is called an *oocyte*. Over the next two weeks, under the influence of FSH, the follicle produces the hormone estrogen, which causes the uterine lining (endometrium) to thicken in preparation for pregnancy.

On about day 14 of the menstrual cycle (day 1 is considered the first day of menstruation), estrogen produced by the growing follicle suddenly increases and signals the pituitary gland to send out a surge of *luteinizing hormone* (LH). In turn, LH triggers the follicle to release its egg, which moves into the fallopian tube and begins its journey to the uterus. It is at this point that the egg can be fertilized by sperm present in the tube. The time between the release of an egg until menstruation begins is referred to as the *luteal phase*. On average, the luteal phase is fourteen days in length.

After ovulation, the empty follicle that is left behind in the ovary begins to produce progesterone. Along with estrogen, progesterone prepares the endometrium for the arrival of a fertilized egg. Under the influence of these two hormones, the uterine lining becomes thicker as its blood vessels and tiny secretory glands increase in number and become more closely crowded together.

If fertilization does not occur, the thickened endometrium eventually breaks down and is shed through the vagina during menstruation. Hormones called *prostaglandins* stimulate this breakdown, as well as the sometimes painful contractions of the uterus that are felt as menstrual cramps. Just before menstruation begins, estrogen and progesterone production abruptly stops. This sharp drop then triggers the pituitary gland to send out FSH, and the menstrual cycle begins again.

The Male Reproductive System

The organs and structures of the man's reproductive system are designed to produce and store sperm before it is delivered to the female reproductive tract. In men, the urinary system overlaps to some extent with the reproductive system, as some of the structures are common to both.

Basic Anatomy

Unlike the female ova, which are present from birth, sperm cells are not produced until a male reaches puberty. As sperm are ejaculated, the male reproductive system continually creates more to replace them. Sperm far outnumber the ova produced by a woman: Whereas an adult female has fewer than half a million eggs in her ovaries, a fertile male releases hundreds of millions of sperm during a single ejaculation.

The male reproductive system comprises the following organs and structures (see Figure 2.2):

- The *testicles*, or *testes*, are two small (about two inches in diameter) organs in which sperm cells are produced. They are located within the *scrotum*, a loose pouch of skin that hangs outside of the man's body, behind the penis. The location of the testes outside of the body cavity ensures the survival of sperm, which can be produced only at a temperature that is slightly below normal body temperature (at about 95 degrees).

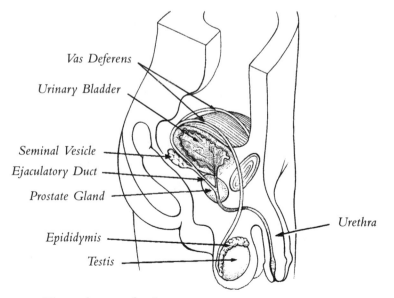

Figure 2.2: The male reproductive system.

- *Sperm cells* are many times smaller than ova. A mature sperm cell is shaped somewhat like a tiny tadpole. Inside the head is the nucleus, which contains the cell's genetic material. Over part of the head is a caplike covering of digestive enzymes designed to break down the outermost layer of the ovum to allow fertilization. The long, whiplike tail of the sperm cell provides it with a means of locomotion.

- Covering part of each testicle is the *epididymis*, the site of continual sperm maturation and storage for mature sperm cells.

- The *vas deferens* (also called the *ductus deferens*) is the long, narrow tube—stretching from the epididymis to the prostate—through which sperm pass on their way out of the body during ejaculation.

- The *seminal vesicles* are small glands located just behind the bladder. They secrete the seminal fluid that helps to lubricate and nourish the sperm cells. This fluid makes up about 60 percent of the volume of ejaculated semen. (Semen is also composed of fluid from the prostate gland.)

- The *prostate gland* lies just below the bladder and secretes fluid as sperm pass through. The *bulbourethral glands* (sometimes called *Cowper's glands*), which lie just below the prostate gland, produce additional fluid to aid the sperm's passage out of the body during ejaculation.

- The male *urethra* serves the dual purpose of transporting both urine and semen. A nerve reflex closes the opening of the bladder and prevents urine from passing through the urethra during ejaculation.

- The *penis* is the male reproductive organ through which semen exits during ejaculation. It is made up of spongy tissue that is tightly packed

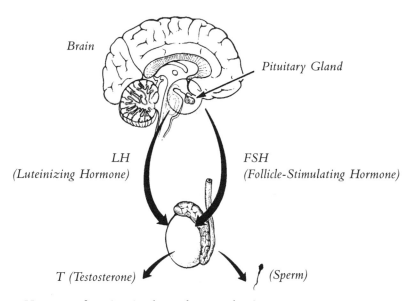

Brain

Pituitary Gland

LH
(Luteinizing Hormone)

FSH
(Follicle-Stimulating Hormone)

T (Testosterone)

(Sperm)

Figure 2.3: Hormone function in the male reproductive system.

with blood vessels. During sexual arousal, the blood vessels in the penis relax and dilate, causing it to stiffen and become erect.

Sperm develop under the influence of some of the same hormones that trigger ovulation in women—including GnRH, FSH, and LH. *Testosterone* from the testes is also uniquely important in sperm development. As in women, the hypothalamus is the central regulator of the reproductive hormones in men. The hypothalamus controls the production and release of GnRH, which is produced in the pituitary. GnRH in turn triggers the release of FSH and LH (see Figure 2.3). These two hormones stimulate the testes to produce sperm. The testes themselves produce testosterone, which helps to maintain the production of sperm.

It takes about two months for sperm cells to mature inside the testes. During sexual arousal, mature sperm cells move from the epididymis, through the vas deferens, and into the urethra as the seminal vesicles and prostate secrete their nourishing fluid. During ejaculation, rhythmic contractions of muscles at the base of the penis send the semen through its opening.

The Process of Pregnancy

Assuming a man and a woman each have healthy reproductive systems and are producing normal sperm and ova, the stage is set for a pregnancy to occur. Fertilization, however, is itself a complex process requiring several more processes to be successfully completed.

Fertilization

About two to five milliliters of semen (about a tenth of an ounce) containing millions of sperm enter the vagina during intercourse. The sperm cells, propelled by the whiplike motions of their tails, then begin to make their way through the cervix and uterus and into the fallopian tubes. On this journey, the sperm face several obstacles that discourage all but relatively few from reaching the egg. The first obstacle is the naturally acidic environment of the vagina, which is hostile to sperm. Once through the lower vagina, the surviving sperm must push their way through the mucus that fills the cervical canal. Of the millions that start the journey, only a few hundred sperm, at most, will reach the fallopian tubes.

It is not entirely clear how sperm "know" which direction to take to reach the egg. It is thought that the motions of their tails, combined with the movement of specialized cells (called *cilia*) in the fallopian tube, aid them in this task. Some sperm reach the ovum very quickly: The journey from the cervix to the ovum is usually completed in four to six minutes.

The sperm that manage to reach the ovum proceed to swarm over it. Usually only one sperm cell, however, will succeed in getting through its surface, using the enzymes on its head (called the *acrosomal cap*) to break down the outer membrane of the egg. Immediately after the first sperm breaks through the egg, a chemical reaction on the ovum's outer membrane usually prevents any other sperm cells from penetrating it.

Development of the Fertilized Ovum

Once a sperm cell is through the ovum's outer membrane, the two cells combine their genetic material, creating the full array of forty-six chromosomes needed to make a new human being. The fertilized ovum—now called a *zygote* or *embryo*—continues its trip to the uterus. As it moves through the fallopian tube, the embryo begins to divide, first into two cells, then four, then eight. Two to three days after fertilization, the embryo is at the sixteen-cell stage, called the *morula* (from the Latin word for "mulberry," for its resemblance to a cluster of berries). By days four to five, the embryo has reached the end of the fallopian tube and has further multiplied into a *blastula*, a two-layered structure of cells surrounding a fluid-filled cavity. The innermost of these two layers is made up of the cells that are destined to grow into a fetus.

Implantation

Upon reaching the uterus, about five to six days after ovulation, the embryo attaches to the endometrium, usually on the upper rear wall of the uterus. During implantation, the embryo secretes enzymes that break down the cells on the surface of the endometrium, allowing it to burrow underneath. The surrounding cells of the endometrium then move over the opening, completely covering the embryo.

The part of the embryo that is buried deepest in the endometrium will eventually develop into the *placenta*—the thick pad of tissue that provides nourishment to and disposes of waste from the growing fetus. The other end of the embryo, the part closest to the uterine cavity, will grow into the fetus and the fetal membranes.

Factors Affecting Fertility

You can see that many conditions must be just right in order for a successful pregnancy to occur:

- The woman must produce a normal, healthy ovum, which must be released into a fallopian tube.

- The fallopian tube must be open to allow the ovum to pass through.

- Intercourse must occur at the appropriate time.

- The man must achieve and maintain an erection long enough to ejaculate semen into the woman's vagina.

- The man must produce enough healthy, actively moving sperm to allow some to survive the trip to the fallopian tubes.

- The sperm must be able to move from the man's testes, where they are produced, out through the penis during ejaculation.

- The sperm must be able to propel themselves through the cervix, uterus, and fallopian tube to reach the ovum.

- The sperm must be able to penetrate the outer membrane of the ovum.

- The embryo must have a clear, unblocked pathway through the fallopian tube into the uterus, and must implant in the endometrium.

A change in any one of these conditions can be enough to prevent conception. Among the many possible causes of infertility related to this process are the following:

- Hormone imbalances that impede or prohibit ovulation, endometrial development, or sperm production

- Anatomical abnormalities that prevent sperm from moving through the male reproductive tract; anatomical abnormalities that impede the ovum's progress through the fallopian tube

- Conditions in the sperm or egg that prevent fertilization

- Impotence in the male

- Conditions of the woman's uterus that prevent implantation of the fertilized ovum

Chapter 4 gives more details about the possible causes of infertility and how your physician may diagnose them. The following section focuses on aspects of age, general health, and lifestyle that might affect fertility.

Age

In some cases of infertility, the woman's age may be a factor. Since no new ova are formed throughout a woman's life, the number of eggs in her ovaries steadily declines as she ages. In her prime reproductive years, a woman has up to 400,000 ova; by age 40, this number has decreased to less than 200,000. By menopause, the number has reached almost zero. More importantly, the quality of the ova also decreases with age. Although a man's sperm production decreases after age 50, this in itself is not a severe enough change to cause infertility.

Over the past few decades, more and more women have delayed pregnancy until their 30s or even their 40s. Delayed childbearing is one factor contributing to the rise in the number of women who consult with infertility physicians.

Sexually Transmitted Diseases

A woman's fertility can be permanently damaged by complications from some *sexually transmitted diseases* (STDs). These infections of the lower genital tract are spread through sexual contact with an infected partner. The rate of STDs has been increasing in the United States over the past few decades, especially in young people between the ages of 15 and 25. Many STDs can be prevented by the use of condoms and by limiting the number of sexual partners.

Two of the most common STDs in the United States are *chlamydia* and *gonorrhea*. These infections often occur together, and they cause no symptoms in as many as half of infected women. Both gonorrhea and chlamydia can be successfully treated with antibiotics. If they are not treated soon enough, however, these STDs may ascend to a woman's upper reproductive tract—the uterus, fallopian tubes, and ovaries—resulting in pelvic inflammatory disease (PID).

PID affects more than three-quarters of a million women each year. When this infection reaches the fallopian tubes, it can leave scar tissue that may partly or completely block the passage of an egg or sperm. If fertilization does occur in a damaged tube, it may result in an *ectopic pregnancy*— a pregnancy in which the fertilized egg implants outside of the uterus, usually in a fallopian tube. Because a fallopian tube cannot expand enough to contain a growing embryo, it will rupture and bleed if the embryo is not removed, posing a serious threat to a woman's life.

In women, the first sign of chlamydia or gonorrhea may be an unusual vaginal discharge; or, if the infection has progressed to PID, a dull ache in the lower abdomen. Because these two STDs occur together so frequently, women are often offered testing for both infections during routine gynecologic exams or other physician visits. In men, the most common symptoms of an active STD are burning during urination and a discharge from the penis. Often, however, men with chlamydia or gonorrhea also have no symptoms. As in the female, untreated STDs in the male can result in obstruction of the duct system.

Medical Conditions

Some chronic health problems might decrease fertility in both women and men. Some of the conditions that can cause infertility in either sex are listed here:

- Blood diseases such as *sickle cell anemia* seem to decrease sperm production in younger males. Other forms of anemia, particularly *beta-thalassemia*—a disease primarily affecting people of Mediterranean descent—are linked to female infertility.

- *Cirrhosis of the liver* is a disease in which liver cells are damaged due to excessive alcohol intake or infections such as hepatitis. This disease seems to be linked to lower sperm levels in adult males, perhaps because of the important role of the liver in metabolizing reproductive hormones.

- Infertility is sometimes caused by poor nutrition, resulting in a shortage of certain vitamins required for the proper release of the sex hormones governing ovulation and sperm development.

- *Inflammatory bowel disease* (IBD), or *Crohn's disease*, is marked by recurrent episodes of diarrhea, fever, and abdominal pain. The causes of

IBD are largely unknown. Gynecologic disorders, including menstrual abnormalities and infertility, are common in women with IBD. In men, some of the drugs used to treat the disease may be linked to decreased sperm production.

- *Hypothyroidism* is a disorder in which the thyroid gland does not produce enough thyroid hormone, important for normal growth and metabolism. In women, hypothyroidism is linked to infertility due to imbalances of the sex hormones responsible for ovulation. Hypothyroidism is a rare cause of male fertility problems.

Lifestyle Choices

Although you can't always change the fact that you have a chronic medical condition, there are things you may be able to do to maximize your chances of conceiving. Couples trying to become pregnant are wise to limit any potential barriers to conception as far as is possible. Both partners can increase their chances of success by eating well, balancing rest and exercise, and reducing the use of alcohol, cigarettes, and any other drugs not prescribed by a doctor.

Nutrition and Exercise If you have been trying to become pregnant, you probably know that being fit and eating right are important both before and during pregnancy. Some nutrition and exercise habits can contribute to infertility in both men and women.

Being extremely over- or underweight can inhibit production of the hormones needed for ovulation and sperm formation. This is especially true when a great deal of weight is gained or lost in a very short period. In both sexes, being underweight can decrease the amount of GnRH sent to the pituitary; and obesity can interfere with the pituitary's release of LH and FSH in response to GnRH, as well as increase the conversion of estrogen to testosterone in the fatty tissues.

Metropolitan Life Height and Weight Table for Women

Height (feet/inches)	Small frame (pounds)	Medium frame (pounds)	Large frame (pounds)
4'10"	102–111	109–121	118–131
4'11"	103–113	111–123	120–134
5'0"	104–115	113–126	122–137
5'1"	106–118	115–129	125–140
5'2"	108–121	118–132	128–143
5'3"	111–124	121–135	131–147
5'4"	114–127	124–138	134–151
5'5"	117–130	127–141	137–155
5'6"	120–133	130–144	140–159
5'7"	123–136	133–147	143–163
5'8"	126–139	136–150	146–167
5'9"	129–142	139–153	149–170
5'10"	132–145	142–156	152–173
5'11"	135–148	145–159	155–176
6'0"	138–151	148–162	158–179

Table reprinted courtesy of Metropolitan Life Insurance Company, Statistical Bulletin. *Weights at ages 25 to 59 based on lowest mortality. Weight in pounds according to frame (in indoor clothing weighing three pounds; shoes with one-inch heels).*

Since being either underweight or obese can interfere with ovulation and sperm production, both women and men wishing to conceive should maintain a healthy body weight—between 80 and 130 percent of their ideal body weight. The chart on page 29 shows ideal body weights for women aged 25 to 59. For example, a 30-year-old woman who is 5 feet 4 inches tall, has a medium frame, and weighs 130 pounds is within her ideal body

Metropolitan Life Height and Weight Table for Men

Height (feet/inches)	Small frame (pounds)	Medium frame (pounds)	Large frame (pounds)
5'2"	108–121	118–132	128–143
5'3"	111–124	121–135	131–147
5'4"	114–127	124–138	134–151
5'5"	117–130	127–141	137–155
5'6"	120–133	130–144	140–159
5'7"	123–136	133–147	143–163
5'8"	126–139	136–150	146–167
5'9"	129–142	139–153	149–170
5'10"	132–145	142–156	152–173
5'11"	135–148	145–159	155–176
6'0"	149–160	157–170	164–188
6'1"	152–164	160–174	168–192
6'2"	155–168	164–178	172–197
6'3"	158–172	167–182	176–202
6'4"	162–176	171–187	181–207

Table reprinted courtesy of Metropolitan Life Insurance Company, Statistical Bulletin.
Weights at ages 25 to 59 based on lowest mortality. Weight in pounds according to frame (in indoor clothing weighing five pounds; shoes with one-inch heels).

weight range of 124 to 138 pounds. To stay healthy, she should not weigh less than 99 pounds or more than 179 pounds.

Exercise is an important part of a healthy lifestyle, but, as in all things, moderation is the key. In women, too much strenuous exercise can cause menstrual periods to become irregular or stop altogether. The absence of periods, known as *amenorrhea*, is common among some professional athletes and dancers. It is important to keep exercising if you are trying to conceive, but it is also a good idea to talk to your physician about how much is enough—and how much might be too much. Male athletes rarely have a fertility problem related to overexercise, but moderation is always preferred.

Alcohol, Tobacco, and Other Drugs Infertility seems to be more common among both women and men who drink to excess than among those who drink in moderation. In both sexes, heavy drinking can interfere with the production and release of sex hormones. Alcoholism can lead to cirrhosis of the liver, which also reduces fertility (see "Medical Conditions," page 27).

Cigarette smoking creates multiple health hazards. It is linked to both low sperm count and miscarriage. There is some evidence that even exposure to secondhand smoke may affect fertility in both sexes. Whether exposure is first- or secondhand, cigarette smoke can impair ovulation and sperm development. Studies show that women who smoke are nearly four times more likely than nonsmokers to take more than a year to become pregnant. Smokers also tend to have more difficult pregnancies.

Recreational street drugs also affect fertility. Women using marijuana are twice as likely to be infertile as those who don't; marijuana use by men may be a factor in infertility as well. Cocaine use has been linked to tubal infertility and low sperm counts. Research into the effects of illicit drugs on fertility continues to uncover new information in this area, but indications are clear that street drugs interfere with fertility.

Environmental Hazards Every day, millions of people are exposed to chemicals that have been added to food, pollutants from cars and factories, and toxic substances in the workplace. Even before birth, a fetus is exposed to many chemicals in the food and air taken in by the mother. Science has yet to determine all of the ways in which these environmental substances affect our health. Some are known, however, to interfere with fertility.

IS MALE INFERTILITY ON THE RISE?

Over the past two decades, several studies have reported apparent declines in semen quality in industrialized nations throughout the world. These reports have suggested that the decline is due to lower sperm counts and higher numbers of abnormal sperm, possibly as a result of environmental toxins. This issue has gained attention among the general population, and speculation has been fueled by the so-called "infertility epidemic"—the increase in the number of couples seeking treatment for infertility in recent years.

Are sperm counts really declining? The answer depends partly on perspective. Many of the studies on this subject have used men donating sperm to sperm banks, attending infertility clinics, or about to have a vasectomy (surgical male sterilization). These groups are not always typical of the general population, so data from them may not apply to the rest of the world.

Although some researchers claim that the decline in semen quality is real, others say there's not enough information to decide this question. They point out that sperm counts can be seen as increasing or decreasing, depending on how the data is statistically analyzed. One study investigated the quality of semen from sperm banks in three different cities, and found that the highest counts were from men in the biggest, most polluted of the three locales. Such a finding seems to contradict the suggestion that environmental pollutants are causing increases in male infertility.

It makes sense for men wishing to conceive a child to limit their exposure to known toxins as much as possible. But until more is known about their effects on sperm quality, and until more men from the general population are studied, it will be hard to draw final conclusions concerning the reality of decreased semen quality.

Diethylstilbestrol (DES) is a drug that was used in some pregnant women in the 1950s, when it was thought to help prevent miscarriage. It was later discovered that DES caused birth defects in some of the babies whose mothers took it during pregnancy. Many of these babies, both female and male, were born with abnormalities of the reproductive organs that have the potential to interfere with fertility. Though some women who were exposed to DES in utero have had difficulty conceiving, DES-exposed males have not been shown to have decreased fertility.

Certain drugs used to treat cancer can damage the ovaries and testes in humans, preventing egg and sperm production. Nitrous oxide, a gas used for anesthesia, can impair fertility in women exposed to high concentrations. A number of prescribed medications have been found to affect male reproduction. Heavy metals, such as lead and mercury, and hydrocarbons, which can cause cell death, are also suspected to impede fertility in workers exposed to them.

High temperatures can prevent the development of sperm, which can survive only at temperatures slightly below normal body temperature. Men should avoid prolonged time in hot tubs and saunas. Similarly, tight clothing can press the testicles against the body, theoretically raising their temperature and inhibiting sperm production.

Couples wishing to conceive should talk to their physician about any over-the-counter or prescribed medications that either partner is taking. A review of the chemicals and other substances encountered at work may help to prevent or limit exposure to those that may affect fertility.

Sexual Habits

Sexual intercourse comes naturally enough to most people. For those wishing to conceive, however, certain key points may be crucial in increasing the odds of pregnancy.

How often a couple has intercourse is an important factor when they are attempting to conceive. Intercourse must take place often enough to maximize the chances of conception. Too-frequent intercourse, however, reduces the amount of sperm in the man's semen, as well as the volume of semen that is ejaculated. So intercourse must take place neither too infrequently nor too often.

Sex must also take place at the right times during a woman's menstrual cycle. The most fertile days in the cycle are the three days before ovulation and the day after. An ovum can be fertilized within twelve to twenty-four hours after it is released, and sperm in the fallopian tube remain able to fertilize an egg for up to forty-eight hours. The likelihood of pregnancy in a given cycle is therefore increased if sex takes place as close as possible to the time of ovulation. For couples trying to conceive, intercourse is recommended every other day around the ovulation peak. Sometimes simply knowing the optimal time for intercourse is all that is needed for a couple to achieve pregnancy. (See "Ovulation Prediction and Timed Intercourse" in Chapter 5.)

Some vaginal lubricants are toxic to sperm. If you want to use a lubricant during sex, check with your doctor to find out which ones to avoid.

Sexual Problems Sexual problems are common causes of infertility and are best dealt with in a straightforward manner. A compassionate and understanding professional can rule out possible underlying causes of many sexual problems. Couples often benefit from therapy or counseling to address the complex issues that sometimes surround sexual problems and infertility.

Impotence If a man cannot achieve or maintain an erection long enough to ejaculate inside the woman's vagina, pregnancy will remain elusive. Impotence can result from some medical conditions, such as disorders of the vascular or nervous system. It can also be a side effect of certain medications, such as those used to treat high blood pressure. Impotence due to medical causes can often be treated by changing the treatment for the underlying problem. Some types of impotence can also be treated with a new medication, sildenafil (Viagra), which became available in 1998 (see Chapter 4).

Psychological stress can cause or exacerbate impotence. Coping with infertility is extremely difficult and trying for many couples. It is not uncommon for either partner to feel pressured by the desire to have children, especially if time to do so is limited by the woman's age. Men in our society have traditionally been discouraged from showing emotion or "cracking under pressure." Keeping their frustration and disappointment inside can exert additional emotional stress and further contribute to impotence. When confronted with impotence, it is essential for men to seek professional help.

Dyspareunia and Vaginismus Some female sexual problems can also contribute to infertility. Dyspareunia, or painful intercourse, can make vaginal penetration difficult or impossible. This condition can result from decreased vaginal lubrication as a woman nears menopause, from pelvic conditions such as endometriosis (in which tissue like that lining the uterus is found in the pelvic cavity), or from infections of the urinary tract or reproductive organs, including PID.

In *vaginismus*, the muscles in and around the vagina undergo painful spasms when vaginal penetration is attempted. Vaginismus may be due to a medical condition but is more often related to psychogenic causes such as fear, or posttraumatic stress from a past sexual assault or injury.

Unexplained Infertility

Even under the best of circumstances—when both partners are healthy and are having sex on a regular basis—pregnancy is not assured. Because the process of fertilization is naturally unpredictable, it is hard to say whether a couple who has not conceived is infertile, or whether simply not enough time has passed. The odds are that with regular, unprotected intercourse, conception will occur in most couples within a year. For many couples, however, it takes longer.

The unfortunate truth is that in up to 10 percent of cases when a healthy couple cannot conceive, no cause can be found. The term *unexplained infertility* does not imply that there is no cause. It simply means that the currently available diagnostic tests cannot identify the reason for a couple's infertility.

Unexplained infertility serves to remind both physicians and patients of the limits of modern medical knowledge. As you encounter the abundance of information that exists in the area of human reproduction, it is important to remember that current diagnostic tests cannot always find a reason for infertility. Though scientists have made astonishing advances in the knowledge of human reproduction, many mysteries remain. For example, a physician can determine how many sperm a man is producing, but many aspects of a sperm cell's ability to fertilize an egg are not completely understood. Likewise, it is possible to detect the hormonal signs of ovulation, but it cannot be confirmed with 100 percent certainty that a normal, healthy ovum is present in a fallopian tube.

As many as one-fifth of couples who have been diagnosed with unexplained infertility do eventually become pregnant. As more research explains more of the mysteries of the human reproductive system, the growing number of options and treatments for infertility will steadily increase and become more successful.

3

Practical Matters and Decisions

Infertility medicine, like the reproductive process itself, is complex. When you begin identifying options in this field, you may feel bewildered by all the choices. The following information is designed to help you make decisions about when to seek infertility care, where to turn for advice, how to choose a physician, and which health care insurance plans will offer the most benefits. To help you get started, this chapter offers some practical pointers on sharpening your general decision-making skills. It also includes examples of questions about infertility you may want to ask along the way. Gathering information and educating yourself about your possible choices are the first steps.

Asking Questions and Making Choices

If you are thinking about seeking infertility diagnosis and treatment, you probably know there is a great deal of information available. But how do you go about finding it? More to the point, how do you even know what questions to ask? And how do you begin to sort through and choose among all your options?

Asking questions and making decisions about your treatment are actually skills—skills that can be learned and improved. If you are feeling confused or overwhelmed, approaching this task step-by-step will help you to make sense of the abundance of infertility information, as well as the available options. And remember that you do not have to commit to anything at this point. Right now, you are just gathering information to see what your options might be.

What to Ask

Making the right choices, of course, means asking the right questions. Some general questions will help to narrow and further define what you're looking for. Here are the questions you may wish to begin with:

- How do you know when it is the right time to consult an infertility expert?

- What type of specialist should you look for?

- How do you go about choosing a physician and/or clinic?

- What credentials should your physician have?

- How do you find out about a physician's/clinic's past performance and successes in infertility treatment?

- How much will various diagnosis and treatment options cost?

- What does your insurance cover?

- How much time and energy are you willing to spend?

As you consider these questions, others will no doubt occur to you as well. Make a list of your questions, and then try ranking them in order of those that concern you most. As you begin to learn about your options, strike off questions that you have found answers for and those that no longer seem relevant, and add new ones as they arise.

Developing Your Decision-Making Skills

Making decisions is a difficult task for many people. Learning how to make rational choices can be enormously helpful when you are faced with what seems like an overwhelming amount of information.

The tips given here apply to almost any type of decision, no matter how large or small. You can apply this technique throughout the infertility process, from choosing a doctor to deciding which treatment options you might try.

Weighing Pros and Cons Most decisions in life have both desirable and undesirable consequences. The trick to making good decisions is knowing how to make the positive results outweigh the negative consequences. A "good" decision is one that leaves you feeling more satisfied than dissatisfied, regardless of the outcome.

To illustrate the decision-making process, we will use the example of Polly and David, both of whom are 34 years old. Married for three years, they had always thought that they would one day have children—but they have been trying for a year and a half without success. They are considering talking to a doctor about the possibility that they have an infertility problem, but they have both heard from close friends that the process of infertility diagnosis and treatment can be time-consuming, costly, and frustrating. Is this a path they want to take? Should they just try a little longer to get pregnant?

Polly and David's delay in getting pregnant causes them both to think more carefully about their wishes for a family. Is having children something they have always just taken for granted, or is it truly important to them?

Polly and David will need to find out what their options are. They will then have to weigh the pros and cons of each choice. To do that, they must first identify their priorities.

Establishing a Strong Foundation We make our best decisions when we first assess the foundation on which we will build once the decisions are made. The cornerstones of that foundation are

- personal values,

- realistic goals, and

- the resources available to achieve those goals.

Polly and David's *personal values* include loving and caring for a child. David feels that a child will enhance not only their own immediate family, but will also strengthen their ties with their parents, siblings, and aunts and uncles. Polly has always wanted a son or daughter to whom she can pass on her love of music. Polly and David both have the *goals* of carrying on their family names and giving their parents a grandchild. Finally, the *resources* they bring to these goals are the ability to support each other emotionally, their support network of family and friends, good physical health, and financial stability.

Examining their values, goals, and resources helps Polly and David to prioritize their decisions. They decide that having children is definitely a priority for both of them. Consequently, the decision to ask Polly's gynecologist for advice is an easy one.

A Step-by-Step Process Breaking down the decision-making process into a series of steps usually makes it seem less perplexing. The steps described here vary according to the choice being made, but generally they can be applied in some way to any decision, large or small.

1. Identify your values, goals, and resources.

2. Learn what the options are, and then examine each one by listing both its positive aspects and its potentially negative consequences relative to your values, goals, and resources.

3. Tentatively choose one of the options. Spend some time imagining the possible outcomes of that decision and how you would feel about each outcome.

4. If necessary, sleep on your preliminary decision.

5. Review your choice and see if it coincides with your intuition.

Here are some general tips that help Polly and David:

- Periodically, they each review their goals, values, and resources, finding that sometimes they change. Their income may go up or down. A major source of stress in their lives may appear or disappear. Or they may decide that something that was once important no longer seems as significant in light of new information.

- Neither tries to make an important decision when tired, upset, or under great stress. When it's 11:00 at night and Polly has an important meeting at work the next morning, she puts off trying to decide which of two doctors she will choose or what to do about her health insurance plan. Rather, she saves those decisions for a time when she is feeling relaxed and rested.

- David tries not to make a big decision in a hurry. Chances are that the more anxious he feels about it, the more important it is. If he thinks he has decided what to do in a given situation, he "tries out" all the possible ramifications in his imagination. Then, he sleeps on it to see whether it still feels right in the morning.

- Decision making is more complex when it affects both parties in a relationship. Polly and David always talk over decisions, but they also allow some "alone time" for each of them to identify individual feelings. They realize that although they make decisions together, they are still two unique people who sometimes differ in their personal values, desires, goals, and resources. Then they come together again and share what they have been thinking and feeling about the decision. They try to compromise on points that are less important than others.

- Polly and David seek advice from a variety of sources. She has an obstetrician/gynecologist she trusts, and she asks whether they should be referred to a subspecialist. At this point, a huge amount of education typically occurs in regard to how fertility is defined, what causes are likely, and the statistics relating to success. David and Polly also talk to friends. Though they realize they can't diagnose their own case based on friends' experiences, talking with people who have had infertility evaluation, diagnosis, and treatment lends moral support and helps them make the decision whether and when to begin the process.

- Finally, Polly and David may eventually find that a mental health counselor is valuable in helping them sort through their decision-making process. A realization that infertility is a couple's issue and not just one person's fault is often emancipating. In approximately one-third of couples, more than one infertility problem is discovered. A mental health counselor would help, not just in making decisions, but also in dealing with the emotions that come up at various points. There are many therapists and counselors who specialize in helping people with infertility problems. Local county or community mental health centers, your local yellow pages, and recommendations from friends or family are good places to look.

For many couples, deciding to see a specialist is a formidable step. Just acknowledging the need for medical advice makes infertility feel like a fact rather than a distant possibility. Anxiety is, understandably, at a peak. At this point, determining your options and educating yourselves about the process will go a long way toward easing anxiety. Education is the key to appropriate decision making. It is also important to obtain your information from reliable sources. For instance, anyone can put information on the Internet, but it may not have been scientifically reviewed. Ask your physician for suggestions about reference material.

Today's Health Care Environment

As Polly and David begin gathering information, they realize that, before they decide what type of doctor to choose and what kinds of procedures to consider, they need to know what their financial resources are. Does their health insurance cover infertility services?

No doubt you have read and heard reports in the news media about the changing face of health care in the United States today. These changes have affected almost everyone with health insurance. They have arisen in part out of changing attitudes about health itself.

More people today are taking a *proactive* stance toward their health and health care. Patients are generally more educated about health matters than in the past. And more people are looking at their health, not just in terms of whether they are sick, but rather in terms of how healthy they are and how they can maintain and improve their health through preventive care. Infertility medicine is no exception to this changing perspective. Many people seeking infertility services are knowledgeable about the process of reproduction and the many treatment choices available.

Such a proactive stance can only enhance the care you receive in today's health care environment. Understanding the basics of your health care arrangements can also optimize your care.

The Meaning of Managed Care

Amid all the ongoing news about reforming the health insurance industry, you may have heard the term *managed care* quite often. If you receive your health insurance through an employer, chances are that you are in some type of managed care plan. What exactly is managed care? And what does this mean in terms of coverage for infertility services?

Managed care began in the early 1900s as a way to coordinate the provision of health care services while controlling costs. In 1973, Congress passed the Federal Health Maintenance Organization (HMO) Act, which spurred the growth of these plans. Today, three-fourths of people who receive their health insurance through their employers are enrolled in some type of managed care plan.

Sorting Out Managed Care Plans Traditional insurance plans are often referred to as *fee-for-service* or *indemnity* plans. In this arrangement, you pay a monthly premium to the insurance company. When you need health care, you choose your own doctors and usually pay for their services as you receive them. You then file a claim with your insurance company to reimburse you for the portion of the doctor's fee—say, 80 percent—that is covered under your plan. In fee-for-service plans, some services, such as preventive care or screening tests, may not be covered.

In a managed care plan, you also pay a monthly fee. Unlike in a fee-for-service plan, you agree to use physicians, hospitals, and other health care providers that have been approved by the plan. Instead of paying a percentage of the cost of services, you pay the physician a copayment— usually 5 to 15 dollars—at the time you receive the services. The rest of the cost is picked up by the insurance company.

Health maintenance organizations (HMOs) are a common type of managed care plan. In most HMOs, you choose a primary physician from a list provided by the organization. In many HMOs, the plan's primary care physicians practice in the plan's own medical center. Many HMO medical centers also employ their own specialists and testing centers, thus providing all care and services under one roof.

Most HMOs use a model called the *individual practice association* (IPA). In an IPA, you choose a primary care physician from a list of those participating in the plan. You see your primary care physician in his or her own office. If you need care from a specialist, your primary care physician refers you to someone on the IPA's approved list.

Preferred provider organizations (PPOs) are another type of managed care. PPOs provide care through a network of physicians and hospitals that give the organization discounts on their usual rates. Like HMOs, PPOs charge patients a monthly fee. In some PPO plans, you may also have a primary care physician, to whom you pay a copayment at the time of your visits. This physician acts as a "gatekeeper" by approving any services received by other specialists. When you need health care, you would contact your primary care physician first, who would then decide whether you need to see a specialist. Specialists are chosen from a list approved by the PPO. In some PPO plans, you may see a doctor outside the plan, but would then pay more costs out of pocket.

Managed care has changed more than the ways in which health care is paid for. It has also increased the emphasis on preventive care, thus focusing on keeping people healthier by avoiding costly health problems. By covering screening and preventive services, managed care has encouraged patients to become active participants in improving and maintaining their health. Unfortunately, to reduce costs, many managed care plans have deemed infertility services "nonessential" and have opted not to cover them.

What Are Your Insurance Options?

If you are unsure about what services your plan covers, talk to your insurance plan administrator. Request a detailed text of what your insurance does and does not cover. If you receive your health insurance from your employer, someone in the personnel or human resources department can help you. The customer service department of the insurance company should also be able to direct you to sources of information about what your plan covers. Some states have laws governing the availability of insurance coverage for infertility services.

Twelve states currently have laws covering some type of treatment for infertility: Arkansas, California, Connecticut, Hawaii, Illinois, Maryland, Massachusetts, Montana, New York, Ohio, Rhode Island, and Texas. The laws in these states vary in the ways they define infertility and infertility treatment, the criteria that patients must meet to qualify for coverage, and the specific types of procedures covered (see Table 3.1).

In California, Connecticut, and Rhode Island, for example, health insurance companies must let employers know that coverage for infertility services is available. However, the law does not require those insurers to provide the coverage themselves. Neither does it require employers to include infertility coverage in their employee health insurance plans.

Providing the claims department of insurance companies with a letter that requests an explanation of benefits may be helpful. Many times, coverage for services can be obtained by discussing the health care contract with managerial staff, instead of the first person who answers the phone. Often,

Table 3.1 State Laws on Insurance Coverage for Infertility Services

State	First year	Mandate to cover	Mandate to offer	Includes IVF coverage?	IVF coverage only
Arkansas	1987				x[1]
California	1989		x	No[2]	
Connecticut	1989		x	Yes	
Hawaii	1987	x			x[3]
Illinois	1991	x		Yes[4]	
Maryland	1985	x			x[5]
Massachusetts	1987	x		Yes	
Montana	1987	x[6]		No[7]	
New York	1990				
Ohio	1991	x[8]		Yes	
Rhode Island	1989	x	x		
Texas	1987				x

[1] *Includes a lifetime maximum benefit of not less than $15,000.*

[2] *Excludes in vitro fertilization (IVF) but covers gamete intrafallopian transfer (GIFT).*

[3] *Provides a one-time-only benefit covering all outpatient expenses arising from IVF.*

[4] *Limits first-time attempts to four oocyte retrievals. If a child is born, two complete oocyte retrievals for a second birth are covered. Businesses with 25 or fewer employees are exempt from having to provide the coverage specified by the law.*

[5] *Businesses with 50 or fewer employees do not have to provide coverage specified by law.*

[6] *Applies to HMOs only; other insurers specifically are exempted from having to provide the coverage.*

[7] *Provides coverage for the "diagnosis and treatment of correctable medical conditions." Does not cover IVF as a corrective treatment.*

[8] *Applies to HMOs only.*

Source: American Society for Reproductive Medicine.

patients are told they do not have coverage for infertility because their policy does not include the services requested, when in fact coverage may be obtained if it is not specifically excluded. Patients are their own best advocates, and persistence pays. Because of the variety of insurers, even differences within a given company exist.

Choosing a Physician

Polly and David have concluded that they have the resources they need to begin seeking infertility diagnosis and treatment. Now they need to decide what type of doctor to see. Can their family physician point them in the right direction? Is Polly's regular gynecologist, whom she knows and trusts, qualified in this area? What are the various medical specialties that deal with infertility? Should they go through the difficult process of finding a new doctor for this problem?

This decision, like most, will involve compromises. Finding and building trust with a new doctor—especially in such an emotional and personal area as reproduction—can be frustrating and time-consuming. On the other hand, perhaps a specialist in this area can help Polly and David optimize their chances of success.

Polly and David are not sure that self-described "experts" in this field possess the experience and knowledge needed to help them make the best choices. So they begin by educating themselves about the types of physicians who specialize in diagnosing and treating infertility. Then they find out about the expertise and qualifications of the physicians they are considering to be sure they are making the best choice.

Types of Fertility Specialists

All medical doctors complete four years of medical school, after which they become interns. Internships may be in a specialized area they have chosen, or may be a rotating internship that will expose them to many areas of medicine. After completing the internship, doctors usually decide upon a specialty, and they move on to a residency program in that field. Before becoming certified in that area, they must pass written and oral exams.

A number of medical specialties deal with infertility. Specialists include urologists, obstetrician/gynecologists, and reproductive endocrinologists. The extent of infertility training required to become certified in these specialties varies.

Urologists Physicians who specialize in urology are qualified to treat disorders of the urinary tract. Urologists who specialize in *andrology* diagnose and treat health problems of the male reproductive system.

Obstetrician/Gynecologists Like all medical specialists, ob/gyns receive four years of training in their chosen field. They then must complete two years of practice experience in obstetrics and gynecology in addition to passing a written and an oral exam. Training in obstetrics and gynecology includes some time—typically, about five to twelve weeks—spent on infertility diagnosis and treatment. Most ob/gyns can give advice about infertility services and are qualified to perform some initial tests. Many have completed additional training in infertility diagnosis and treatment and may have extensive experience in this area.

Reproductive Endocrinologists The field of reproductive endocrinology is one of the fastest-growing and most complex areas of medicine. Physicians in this field have had extensive and rigorous training. They draw upon their broad knowledge to help you make the best possible decisions to diagnose and treat infertility.

Requirements for Certification There are about 500 board-certified (and a further 800 board-eligible) reproductive endocrinologists in the United States. To become board certified, a physician must first complete all the requirements for board certification in obstetrics and gynecology, and pass written and oral exams. In addition, he or she must undergo two to three years of training in a fellowship program that is accredited by the American Board of Obstetrics and Gynecology (ABOG). A written examination and two years of practice experience precede a three-hour oral examination in reproductive endocrinology. While completing the practicum, but before taking the oral exam, a physician is board eligible in reproductive endocrinology.

Where to Find a Reproductive Endocrinologist If you are seeking a reproductive endocrinologist, ask your potential doctor whether he or she has been board certified (or is board eligible) by the ABOG in this subspecialty. The doctor may also belong to one or more professional societies, such as the American Society for Reproductive Medicine (ASRM), the American College of Obstetricians and Gynecologists (ACOG), the Society of Reproductive Endocrinologists (SRE), or the Society for Assisted Reproductive Technology (SART). Membership in these organizations requires that a physician meet certain criteria, but is not the equivalent of board certification. A good source of information about board-certified reproductive endocrinologists in your area is the *Directory of Medical Specialists*, published by *Who's Who*, which is available in most public libraries.

🌿 FERTILITY CLINICS AND SUCCESS RATES

There are at least 280 fertility clinics in the United States offering assistance to couples wishing to achieve pregnancy. The procedures offered at these clinics include fertility drugs, surgery, artificial insemination, and *in vitro fertilization* (IVF)— the transfer of fertilized eggs into a woman's uterus.

As you explore the possibility of using a fertility clinic's services, you will of course want to know how successful they have been in helping couples solve their infertility problems. Ask for a clinic's published success rates, and look at their answers carefully. There are several ways in which to measure success rates, and each gives a slightly different picture about the effectiveness of the procedures.

Here are five widely used ways to calculate a clinic's success rate:

1. LIVE BIRTHS PER CYCLE: Sometimes called the *take-home baby rate*, this rate is the percentage of ovulatory cycles that resulted in the birth of a living, healthy child or children (multiples are counted as one pregnancy). This rate is lower than other success measures because some pregnancies end in miscarriage or stillbirth.

2. PREGNANCIES PER CYCLE: This rate is the percentage of ovulatory cycles resulting in clinical pregnancies (defined as the presence of an embryo on ultrasound). This rate is always higher than the rate of live births per cycle, because not all pregnancies are carried to term or result in a live infant.

3. LIVE BIRTHS PER EGG RETRIEVAL: This is the number of live, healthy births resulting from procedures in which eggs are obtained from the woman and used for IVF or another method of infertility treatment. This rate does not count the number of cycles in which no eggs were retrieved (for reasons such as too few follicles or illness in the woman), and so it is usually higher than the rate of live births per cycle.

4. LIVE BIRTHS PER EMBRYO TRANSFER: This rate is the number of live births that resulted from embryo transfer, in which an egg that was fertilized outside the woman's

body is transferred back to the woman. This rate generally yields a higher success rate because it excludes cycles
in which no egg was transferred (for reasons such
as unsuccessful fertilization of the egg, or abnormal
embryos).

5. PREGNANCIES PER EMBRYO TRANSFER: This is the number
of times a pregnancy test is positive after the embryo
transfer. Since this method of calculation excludes cycles
that were canceled cycles, with no embryo transfer, and
includes failed pregnancies, the rate is artificially high.

You should know that the success rates of infertility
treatments are never extremely high. On average, the rate
of live births per cycle in fertility clinics nationwide was
about 20 percent in 1995. Even using a liberal definition of
success, such as the number of live births per embryo transfer, the average success rate was only about 25 percent.

Do not take any clinic's "success" rate at face value.
Look at it carefully and ask how it was calculated. Does the
rate apply to all patients, or only to those under a certain
age? Does it reflect live births only, or all pregnancies? Is it
based on the number of all cycles, or only on those cycles
that were completed? What was the cycle cancellation rate
of the program? Depending on the answers to these questions, a rate that sounds too good to be true may turn out
to be exactly that. Remember, these statistics can be veri-
fied by the clinic-specific reporting (see page 53).

Fertility Clinics Many reproductive endocrinologists belong to group practices in fertility clinics. These facilities provide many of the services needed

by infertile couples, including diagnostic tests and treatment procedures. They usually have a host of experts on staff, including physicians, nurses, and technicians.

Fertility clinics exist in more than half of the states. Most are located on the East Coast and in Southern California. The procedures they offer vary, as do their success rates.

Before deciding to work with the staff of a fertility clinic, gather information about the clinic's experience and successes. A "success" rate can be defined in different ways, depending on how it is calculated. Be careful not to be misled by a clinic's claims of extremely high success rates. As in everything else connected to infertility diagnosis and treatment, asking the right questions in this area is critical. Does the rate reflect the number of ovulatory cycles that led to pregnancies carried to term and resulted in live, healthy infants? This is the number most people are interested in, and it is likely to be lower than other measures of success.

In 1992, Congress passed the Fertility Clinic Success Rate and Certification Act. This law requires the Centers for Disease Control and Prevention (CDC) to publish the success rates of clinics throughout the United States where assisted reproductive techniques (ART) are performed. The first of these reports to be published so far contains data for 1995 in three volumes for different regions. It is intended for laypersons who are considering using assisted reproduction to achieve pregnancy. The report, entitled *1995 Assisted Reproductive Technology Success Rates: National Summary and Fertility Clinic Reports*, was published in December 1997 by the United States Department of Health and Human Services, and is available free of charge. Call (888) 299-1585, or write to RESOLVE, Department ART, 1310 Broadway, Somerville,

MA 02144-1779; or log on at http://www.resolve.org. The
Society for Assisted Reproductive Technology (SART) also has
tabulated clinic-specific rates for all its member clinics. The
information is available from the American Society for
Reproductive Medicine (ASRM), 1209 Montgomery Highway,
Birmingham, AL 35216-2809; phone (205) 978-5000; website
at http://www.asrm.org.

Your Relationship with Your Physician

Once Polly and David have a name or two, they make appointments to talk
with the physicians who seem to be likely choices. They know that the
process of infertility diagnosis and treatment can be emotionally frustrating at times. It is important to them that their doctor is someone they feel
they can trust and rely on. They hope that meeting with the doctor before
beginning a lengthy course of tests and evaluations will help to give them
an idea of whether they would make a good team.

It may seem unusual or inappropriate to make an appointment with a doctor
just to talk. But this is an important decision for you. A doctor who is willing to
give the time to address your questions and concerns is likely to be available
when you are confused or need help in sorting out an important decision.

Before the visit, Polly and David make a list of questions about the physician's practice and philosophy. Following are some of the questions on their
list. When you reach this stage in the process, you will no doubt have more
questions of your own.

- What is the physician's experience with diagnosing and treating infertility? Is he or she board certified in reproductive endocrinology?

- What are the physician's fees for office visits? What payment plans are
  available?

- What are office policies and procedures regarding appointment cancellations and filing of insurance claims?

- Will the physician, or someone else on staff, be available twenty-four hours a day, seven days a week, including weekends and holidays? Are lab and ultrasound facilities open seven days a week?

- If it is a group practice, will they always be seeing the same physician?

- Does the physician make referrals to a urologist who specializes in reproduction issues for evaluation of male infertility?

- What can they expect at their first few visits? What tests will be performed, and what are they expected to do beforehand?

When Polly and David have made their choice of physician, their doctor meets with them to go over what tests and/or treatment they may already have received. This initial appointment gives them an opportunity to ask remaining questions and to form the basis for a good doctor–patient relationship.

As an individual or a couple facing infertility, only you can know your most important wishes, priorities, and goals. Keep those things in mind as you go through this process, and make sure your doctor is aware of them, too.

Participating in the Process

The desire to become a parent—and the difficulty you have had in doing so—is a deeply emotional issue. It probably seems to you that getting pregnant should not have to require so much concentration and deliberation. Beginning the process of infertility diagnosis and treatment, however, is best done with eyes wide open to the decisions that must be made and the possible consequences of those decisions.

You will not be alone. You will be participating on a team whose members all have the same goal. By researching the resources in your area, you can assemble the team that will work best with you, and avoid many pitfalls as the process unfolds. Consider your physician the team leader, responsible for providing treatment options that are medically sound and ordered with regard to efficacy, invasiveness, safety, expense, and convenience. Ultimately, however, the patient is in control of the treatment that is chosen.

Sometimes a fresh look at your medical history reveals something that was previously overlooked. No physician should be offended by a request for a second opinion.

4

Diagnosing Infertility

The great strides taken in reproductive medicine over the last few decades have made it possible to uncover previously mysterious causes of infertility. Diagnosing infertility, however, can still be time-consuming and laborious, and the procedures recommended by specialists may seem as mysterious as the problem itself. This chapter describes the tests and procedures commonly used to diagnose infertility, so that you can begin the process with a working knowledge of what to expect. The tests used and the order in which they are performed will vary with each particular case, but reading this chapter will help you to be prepared to discuss the tests with your physician once they are proposed.

Keep in mind that there are many possible outcomes of this process. Infertility problems are found about equally in men or women in up to 80 percent of cases. About 15 percent of cases are due to combined male and female factors. Some 5 to 10 percent of infertility cases are diagnosed as "unexplained"—no cause can be found. No matter what medical information the infertility workup uncovers, it is bound to raise personal issues for both partners. Chapter 6 explores some of the common feelings and conflicts experienced by men and women going through the process of trying to overcome infertility.

The Infertility Workup

Unless there is already a known reason to do otherwise, the process of diagnosing infertility starts with the simplest tests. Depending on the results of these first evaluations, one or more of the tests described here may be recommended. With a general idea of what these tests entail, you can talk knowledgeably with your infertility specialist about how the infertility workup will proceed. Once a plan is in place for your particular needs, you can refer back to the information here for an overview of what to expect.

Planning the Infertility Evaluation

The specifics of the infertility evaluation vary from couple to couple and depend on many factors, including motivation, resources, and medical histories. Some basic principles, however, will guide the process.

- **Diagnosing infertility will take time.** How much time depends on the particulars of your case. The time required may be anywhere from a month to a year, or even longer in some cases.

- **Both partners must be evaluated.** Infertility is neither a "woman's problem" nor a "man's problem." Successful diagnosis and treatment requires that both partners be motivated to participate.

- **It is best to have a clear plan.** Ask your doctor to spell out clearly the recommended procedures and the order in which they will be performed. Write it all down so you can refer to it throughout the process, and ask questions about anything you do not understand.

- **The evaluation must be complete and thorough.** Many cases of infertility are due to more than one factor. Even when a significant contributing factor is found at some point during the diagnostic process, it is a good idea to complete the tests as planned to ensure that treatment decisions can be made on the basis of all the available information.

- **Questions are crucial.** The diagnostic process affords you a chance to develop a fruitful working relationship with your doctor. It should also give you an opportunity to clear up any confusion or questions you may have. Don't be afraid to ask your doctor about anything you do not understand. And don't hesitate to find another care provider if you feel that your concerns are not being taken seriously.

First Steps: History and Physical

Diagnosing infertility begins with a medical history and physical examination of both partners. The information from this part of the process will help to guide decisions about further tests and treatment.

Medical History The medical history consists of a series of questions each partner will answer individually. The questions themselves and the order in which they are asked vary among doctors and patients, but there are some standard areas of inquiry for both partners. These include:

- The age of each partner

- How long the couple has been trying to get pregnant

- Past difficulties for either partner in achieving conception

- Any previous evaluation or treatment for infertility

- Any earlier pregnancies with either a past or the current partner, the age when those pregnancies occurred, and the outcome

- Frequency and timing of intercourse

- Use of lubricants during sex

- Use of alcohol, tobacco, and other drugs, including prescribed medications, over-the-counter drugs, and "recreational" or illicit drugs

- Exposure to environmental toxins at work, from a hobby, or elsewhere

- Exposure to DES before birth (see Chapter 2, page 33)

Some of these questions may be asked of both partners while they are together in the doctor's office, whereas others, such as those listed below, may be asked during each partner's individual exam.

The man may be asked:

- Ages at which milestones of puberty first occurred (such as appearance of pubic hair, voice change, and first nighttime ejaculation)

- Factors that can impede sperm production

  - Past illnesses and infections, such as mumps (as an adult), cancer of the testes that required radiation treatment, or a recent high fever

  - Use of hot tubs and saunas

  - Drug use, including certain antibiotics, steroids, and illicit drugs such as marijuana and cocaine

- Past injuries or surgery on the pelvic or genital area

The woman may be asked:

- Age

- Age when menstrual periods began

- Ages at which other milestones of puberty occurred (such as breast development and pubic hair)

- Frequency of periods and quality of menstrual flow (heavy, moderate, or light), or changes in menses

- Past or current symptoms of abdominal or pelvic pain, or pain during intercourse

- Past sexually transmitted diseases (STDs), especially pelvic inflammatory disease

- Previous abdominal surgery

- Exercise (how often and how long)

- Previous pregnancies and their outcomes

- History of birth control use

- Drug use, including certain antibiotics, steroids, and illicit drugs such as marijuana and cocaine

Both partners should plan to be forthcoming about all aspects of their medical history. Even if it seems as though a past condition, illness, treatment, or habit could have no bearing on the ability to conceive, it is important to inform your doctor about it.

Physical Exam Information from the physical exam can provide further clues about aspects of the couple's health that may play a role in infertility.

In general, your doctor is looking for abnormalities of both the male and female reproductive systems that might impede pregnancy.

Other clues can be provided by abnormalities in physical development. Improper development of secondary sexual characteristics—for example, decreased body hair or enlarged breasts in the man, or excessive hair growth in the woman—may indicate a hormone abnormality that can impede sperm production or ovulation.

Physical Exam of the Man After the specialist evaluates the male partner in terms of his overall physical development, he or she will examine his reproductive organs. The doctor will look for any abnormalities in the shape or size of the penis or in the location of the penile opening. An abnormality can result in semen being placed in the vagina during sex in such a way that it is harder for sperm to find their way through the cervix. A manual exam of the testes can indicate whether any abnormal growths or lesions are present and whether the testicles have descended properly (see "Cryptorchidism" under "Male Infertility," page 90).

Physical Exam of the Woman For the woman, a pelvic exam gives the doctor information about the size, shape, and position of the uterus and ovaries. It can also uncover the presence of abnormal growths on the ovaries or in the uterus. Tenderness or pain in the lower abdomen can indicate disease or infection. A general physical and breast exam are also usually done.

Laboratory Tests

The information from the history and physical exam is used to guide decisions about what diagnostic tests to perform. These tests generally begin with the less costly and invasive ones and progress to more complicated procedures as needed. Initial lab tests can yield important information about ovulation, sperm production, and the interaction between sperm and cervical mucus.

Tests for Ovulation There are a number of ways to find out whether the woman is producing an egg each month. Because ovulation cannot be observed directly, all of the methods used to detect it are based on looking for signs that are known to accompany the event. For instance, a woman's body temperature and the level of certain hormones change noticeably around the time of ovulation. Changes also take place in the endometrium (the lining of the uterus) and in the mucus produced by the cervix.

Basal Body Temperature The simplest technique for detecting ovulation is to chart the woman's *basal body temperature* (BBT). Each month, just after a woman ovulates, her BBT increases slightly—about 0.6 to 1 degree Fahrenheit. This tiny increase can be detected with a special thermometer that is marked in smaller increments than a regular thermometer. To complete the test, each day throughout one complete menstrual cycle, the woman takes her temperature when she first wakes up in the morning (before beginning any physical activity that would raise her body temperature) and records the findings on a chart provided by her physician (see Figure 4.1 for a sample chart). From her notations, the doctor can see if there is a pattern of increased BBT that determines that ovulation has occurred.

Hormone Tests Results of the BBT method are sometimes less than clear-cut. Even when there is a pattern of BBT surges, many doctors choose to confirm ovulation by measuring a woman's levels of reproductive hormones.

A woman's body secretes the hormone progesterone in increased amounts after ovulation to prepare the uterine lining for a possible pregnancy (see page 65). An increase in progesterone one week after ovulation is believed to have occurred (according to whatever method is used to predict it) is considered to be a reliable sign that ovulation has indeed occurred. Another test requires a kit (sold over the counter in drug stores) that measures levels of luteinizing hormone (LH), which increases sharply at ovulation. Other over-the-counter kits measure levels of progesterone. Your

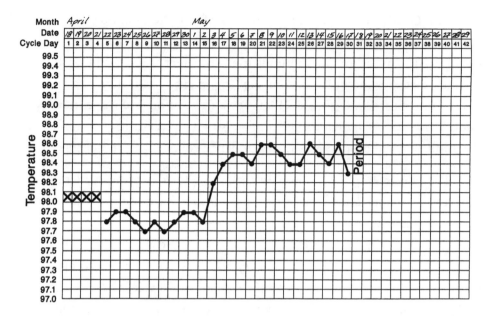

Figure 4.1: Basal body temperature chart showing ovulation. Look for a slight rise in temperature about 14 days after period begins.

doctor may instruct you to use one of these kits at a specific time during the menstrual cycle and to bring the results to his or her office.

Additional hormone tests may be done if there is reason to suspect that the woman has a hormone imbalance that is interfering with her ability to become pregnant. Abnormal levels of *thyroid-stimulating hormone* (TSH) can indicate a thyroid disorder that may be preventing ovulation.

Inadequate thyroid hormone can also be a cause of excess prolactin production. *Prolactin* is a hormone that triggers milk production during pregnancy. In a nonpregnant woman, high prolactin levels can cause several types of reproductive problems, including inadequate progesterone production during the luteal phase after ovulation (see Chapter 2, page 18), irregular ovulation and menstruation, amenorrhea (absence of menstruation), and *galactorrhea* (production of breast milk by a woman who is not pregnant or nursing). Excess prolactin can be caused by certain types of medications (in particular, some types of tranquilizers, high blood pressure medications, and antinausea drugs), by chronic kidney disease, or by abnormal growths on the pituitary gland. Medication to treat excess prolactin can restore hormone balance, ultimately allowing pregnancy to occur.

Endometrial Biopsy While it is a more invasive procedure, the endometrial biopsy can provide information about the function of the woman's reproductive hormones. This minor procedure is done ten to twelve days after the time that ovulation is thought to have occurred. It is usually done in the doctor's office.

A thin, flexible catheter is inserted into the vagina and through the cervix, and a piece of the endometrium is collected (see Figure 4.2). Because of progesterone's effects on the endometrium after ovulation, the appearance of this tissue under a microscope can indicate whether progesterone levels have risen in response to ovulation, and whether the endometrium is developed sufficiently to allow implantation and growth of an embryo.

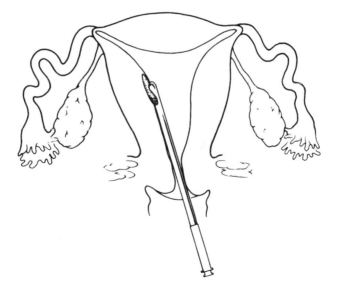

Figure 4.2: Endometrial biopsy.

Tests for Male Infertility Tests and procedures performed on the man focus on identifying problems with the production of sperm or with the ability of sperm to reach the woman's fallopian tubes and to fertilize the egg. A man may undertake these tests at the same time his partner is having her infertility tests.

Semen Analysis Tests to evaluate the man's semen provide important information about his reproductive hormones, sperm quality, and the health of his reproductive tract. To ensure the most accurate results, the man will be asked to refrain from ejaculating for two to three days before the semen analysis. He will then collect a sample of his semen, either by masturbating or by using a nonlubricated condom during sex.

It is best to collect the sample at the office because of better quality control, sperm survival, and pregnancy rates. If the man collects the sample at home, it must be kept at body temperature, and delivered to the lab within forty-five minutes. Many doctors prefer to obtain at least three semen analyses from three different samples.

A number of characteristics can be evaluated in the semen analysis. The first and simplest are the *sperm count*, an assessment of the number of sperm present in each milliliter of semen; and *volume measurement*, how many milliliters of semen are collected in the ejaculate. A low sperm count reduces the potential for a healthy sperm to fertilize an egg. The *motility* (swimming movement) of the sperm is thought to play a key role in their ability to travel through the fallopian tubes and to penetrate the egg.

Technicians evaluate the sperm *morphology*—the physical appearance of sperm cells—by placing a drop of semen on a glass slide and staining it with special dyes (see Figure 4.3). Normal sperm have oval heads and relatively long, whiplike tails. Some abnormally shaped sperm are present in any sample. Abnormal sperm are not thought to cause birth defects. In fact, it is thought that these abnormal sperm are not able to fertilize an egg. However, too many abnormally shaped sperm can be a cause of infertility.

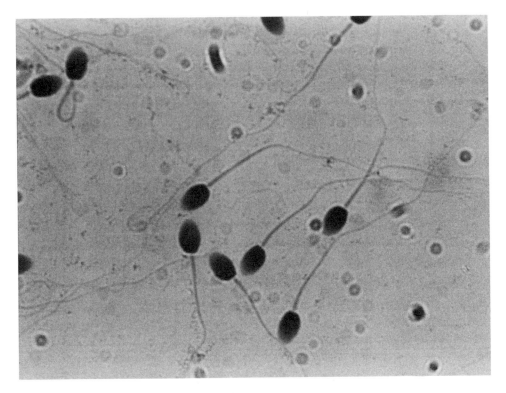

Figure 4.3: Sperm as seen under a microscope.

During the same analysis, the laboratory technicians evaluate the seminal fluid itself. Normal semen coagulates (thickens) after ejaculation into a jellylike consistency and then, over the course of about thirty to forty minutes, becomes liquid again. If this process, called *liquefaction*, does not occur, the sperm might not reach the cervix.

The volume of ejaculated semen is also measured. Seminal fluid is naturally alkaline, or basic, as opposed to the naturally acidic environment of the vagina. Too little protective seminal fluid can make it tougher for sperm to survive the acidic environment of the vagina.

In addition to the consistency and volume of the semen, certain substances in the semen may also be analyzed. The absence of fructose, a type of sugar normally found in semen, may indicate an obstruction of the duct system. The presence of many white blood cells, or leukocytes, is thought to be linked to male infertility, and may indicate infection. If this is the case, antibiotics may improve the man's fertility potential.

Sperm Antibodies Antibodies to sperm are found in about 10 percent of men and in slightly fewer women who are tested for infertility. Antibodies are specialized proteins that are normally produced in the bloodstream in response to an invading organism or infection. Their production is triggered by substances called *antigens*, which are carried on the surface of an invading organism. The job of antibodies is to protect the body against disease by binding to antigens. Each type of antibody fits a specific type of antigen, like a key in a lock. Once attached to the antigen, the antibodies destroy or weaken it so that it no longer poses a threat.

For reasons not completely understood, some types of antigens are normally present in the body—such as on sperm cells. In some men, antibodies against these sperm antigens are found within the reproductive tract. In women, sperm antibodies are found in serum—the liquid portion of blood that remains after the solid components have been removed—and in cervical mucus. Having sperm antibodies does not seem to affect a person's overall health. But sperm that are bound to these antibodies seem to have a diminished ability to swim through cervical mucus or to fertilize an egg.

Sperm antibodies can be found through a test using specially coated microscopic beads. Blood and/or semen samples are collected for this purpose. The reproductive specialist may also call for various other types of tests used to detect sperm antibodies, depending on his or her preference and available resources.

Sperm antibodies are more common in men who have had vasectomy (surgical sterilization), vasectomy reversal, sexually transmitted diseases, or injury to the testes. Men in whom fewer than 20 percent of sperm are bound to sperm antibodies appear to have the same chances of fathering a child as men without sperm antibodies. Higher numbers of sperm antibodies can make fertilization more difficult.

Sperm Penetration Assay (SPA) The sperm penetration assay (SPA) yields information about sperm's ability to fertilize an egg. This test is usually done in couples for whom no other cause of infertility can be found.

In most species of animals, including humans, ova cannot be penetrated by sperm from a different species. The outer membrane of the egg, called the *zona pellucida*, appears to be the key in preventing this occurrence. Even when the outer membrane is removed, the eggs of most species still do not allow penetration by other species.

Eggs from a mature female hamster, however, can be penetrated by healthy human sperm once the zona pellucida is removed. It was this discovery that led to the development of the SPA, sometimes also called the *zona-free hamster ova penetration test.*

In the SPA, ova from a female hamster are stripped of their outer membranes. The sperm collected from the male (human) partner are processed in order to expose the enzymes carried within the cap covering the head (see Chapter 2, page 20 for a description of sperm structure). One method involves mixing the ova and sperm and placing them in an incubator for several hours. Another method uses a special device in which one end of tiny tubes contain the ova while the other end is placed into a container holding the sperm, allowing them to swim toward the ova. Whichever method is used, the ova are then viewed under a special high-power microscope to determine whether any have been penetrated by the sperm.

The SPA is a subject of some debate among fertility experts. There have been problems in setting standards to determine what constitutes a positive or normal result, and the way in which the procedure is done varies among different institutions. A good result from the SPA seems to indicate that a man's sperm is healthy and active enough to successfully fertilize a woman's egg. The SPA, used with other tests of semen quality, can provide valuable information about the fertilizing capacity of a man's sperm. If your doctor suggests this test, you will need to talk with him or her about the details of how it will be performed and what the results mean.

Postcoital Test   The *postcoital test* (PCT) can be used to evaluate the interaction between the man's sperm and the woman's cervical mucus. The secretions normally found within the cervix change in texture and consistency throughout a woman's menstrual cycle. At the time of ovulation, the mucus becomes clear, slippery, and elastic, in order to facilitate sperm motility.

As part of the postcoital test, the woman's cycle is first monitored to identify when ovulation is about to occur. This may be done with one of the ovulation predictor kits that are available over the counter, with BBT monitoring, or with simple observation of cervical mucus.

A reliable method for predicting ovulation, in preparation for the postcoital test, involves the use of vaginal *ultrasound*. This is a procedure in which a transducer, a small hand-held instrument somewhat like a microphone, emits harmless and painless sound waves that are bounced off the structures inside the woman's abdomen and transformed into images that can be viewed on a monitor. With ultrasound, the size of the ovarian follicle can be measured two to three days before the day on which ovulation is expected to occur. Because ovulation is known to occur when the follicle reaches a certain size (about seventeen to twenty-five millimeters in diameter), the PCT can be scheduled to coincide with ovulation.

Once ovulation can be more or less accurately predicted, the couple refrains from intercourse for two days before the PCT is to be done. When signs indicate that ovulation is about to occur, the couple has intercourse, and the woman comes in to the doctor's office six to ten hours later. At that time, a sample of cervical mucus is obtained in a manner similar to a Pap test, and examined under a microscope.

Because the quality of the cervical mucus changes throughout a woman's menstrual cycle, it is possible that at the time of ovulation it may have characteristics that make it difficult for the sperm to move through it. It may be too thick, for instance, or have a chemical makeup that is hostile to sperm. The PCT allows the doctor to assess the interaction between the sperm and the cervical mucus. It indicates whether the sperm are active and healthy enough to move through the mucus at the time the woman is ovulating, when the chances of pregnancy are highest.

Like the SPA, the PCT is also the subject of some debate. There have been no definite guidelines set forth to define what the results mean, and so the outcome of this test is interpreted in different ways among different facilities. If the results of the initial test are unclear, a second exam may be done, this time with the couple coming in two to three hours after intercourse.

Timing is of the utmost importance in the PCT. If examined too soon in the woman's cycle, sperm may appear active and healthy but may perish later as the mucus undergoes its normal changes at ovulation. If the timing is accurate and the results appear good, the PCT can help to rule out a cervical factor as a cause of the couple's infertility (see "Cervical Factors," page 83).

Diagnostic Procedures

After the tests described above—when ovulation has been assessed in the woman, semen and sperm quality have been evaluated in the man, and a PCT has been considered—more complicated procedures may be in order. These procedures are done to identify abnormalities in the woman's reproductive tract.

Depending on a variety of factors, including who is to perform the procedure and the available resources of your infertility clinic, these operations may be performed in a doctor's office or in a hospital. If you are scheduled for a surgical procedure, ask your doctor for details of where it will be performed and whether a hospital stay will be involved.

The side effects of all of these tests are usually minor, but as in any operation, problems can arise. After having any of these procedures, it is important to notify your doctor if you have any of the following symptoms:

- Fever

- Severe pain in the abdomen

- Heavy bleeding or discharge from the vagina

- Fainting or dizziness

- Severe vomiting

Hysterosalpingogram During a hysterosalpingogram (HSG), a special dye that can be seen on an x-ray is introduced through the cervix until it fills up the uterus and fallopian tubes. HSG yields valuable information about the anatomy of these structures—specifically, the size and shape of the uterine cavity and whether there is any obstruction of the fallopian tubes. The information gained from HSG will guide decisions about further diagnostic procedures.

The Procedure Before HSG is performed, the physician will do a pelvic exam to check for masses or tenderness. If he or she suspects pelvic inflammatory disease (PID), the test will be postponed until such a condition can be treated.

The patient remains awake and aware throughout the procedure. A speculum is inserted into the vagina to hold its walls apart and to allow the doctor to see the cervix clearly. After the cervix is swabbed with a cleansing agent, a long, thin instrument called a tenaculum is inserted into the vagina. An edge of the cervix is gently grasped with the tenaculum to hold it in place, and the speculum is then removed.

Next, a syringe filled with a special dye, called contrast medium, is attached to a cannula, a long, thin tube with a small, acorn-shaped tip. The tip of the cannula is pressed up against the cervical opening, and the contrast medium is slowly injected until it completely fills the inside of the uterus and the fallopian tubes.

x-rays taken during the procedure allow the specialist to watch the medium as it moves through the upper reproductive tract (see Figure 4.4a). From these x-rays, the doctor can see the size and shape of the uterine cavity and whether there are any growths inside the uterus or obstructions of the fallopian tubes (as shown in Figure 4.4b).

After the procedure, the contrast medium will eventually dissolve inside the abdomen. A final x-ray may be taken a few hours afterward to ensure that the dye has dispersed properly.

Possible Complications The main side effect of HSG is the pain and discomfort associated with the procedure. The use of a nonsteroidal anti-inflammatory drug (NSAID—such as ibuprofen, naproxen, or aspirin) or a local anesthetic can minimize such discomfort. Other risks of the procedure occur rarely. They include the risk of worsening a previously undiagnosed pelvic infection, and a very slight risk from the x-rays taken during the procedure. A minor risk is also linked to the contrast medium. In rare cases, this medium may find its way into the bloodstream or lymph system. When this occurs, only about 20 percent of women experience side effects, which include temporary chest pain, heartburn, cough, and headache. Serious complications are extremely rare.

Laparoscopy Through a minor surgical procedure called laparoscopy, specialists gain additional information about a woman's reproductive organs. Unlike HSG, laparoscopy allows the doctor to view the internal organs and structures directly, using instruments inserted through a tiny incision in the abdomen. Laparoscopy can be used as both a diagnostic and a treatment

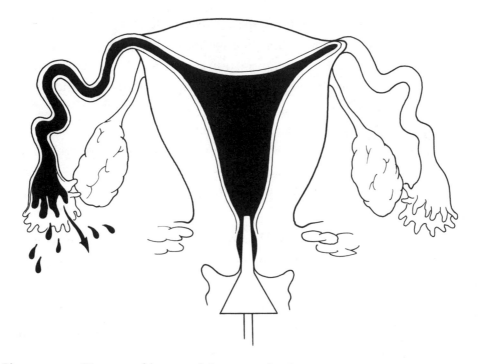

Figure 4.4a: Diagram of hysterosalpingogram (HSG).

procedure, because problems found during the time of the surgery can often be treated at the same time.

In some cases, laparoscopy is scheduled when HSG reveals no problems. (For diagnosis only, a mini-laparoscopy can be done in the office, if the physician has the appropriate equipment.) However, if the HSG shows uterine fibroids (benign growths on the uterine wall) or adhesions (scar tissue that binds structures together inside the pelvis), and if the problem is not too extensive, the laparoscope is used to treat it.

General anesthesia is used for this procedure, which means it is usually performed in a hospital. During general anesthesia, the patient is asleep, feels no pain or discomfort, and is not aware of her surroundings. In most cases, the patient can go home on the day of her surgery.

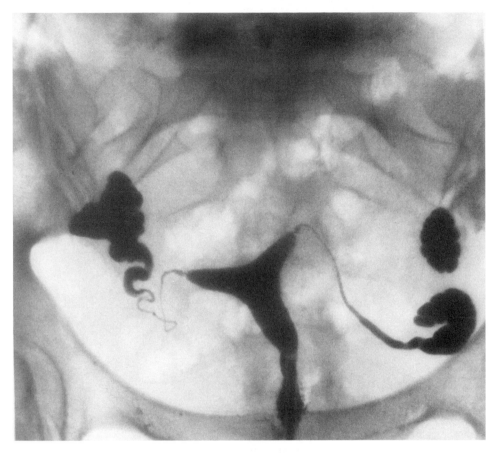

Figure 4.4b: Actual x-ray of hysterosalpingogram (HSG).

The Procedure After induction of general anesthesia, the patient is positioned with her feet in stirrups and her hips slightly elevated to allow the bowel to shift upward and out of the pelvis. A tiny incision is made in the abdomen, usually through or just below the navel, through which the instruments are to be inserted. A gas such as nitrous oxide or carbon dioxide is then introduced into the abdomen. This expands the space inside the abdomen to create more room in which the surgeon can maneuver, and pushes the abdominal wall away from the internal organs.

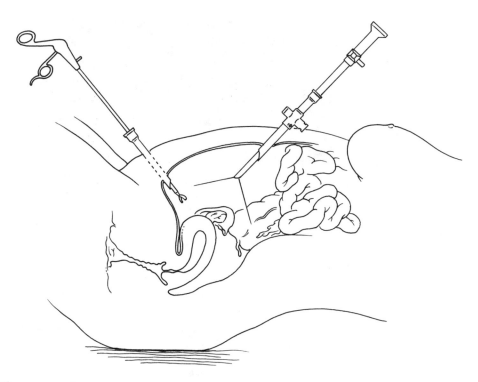

Figure 4.5: Laparoscopy.

The laparoscope is a long, slender instrument with a light and a magnifying eyepiece on one end, somewhat like a miniature telescope. It is inserted through the abdominal incision so that the surgeon can directly view the uterus, tubes, and ovaries (Figure 4.5). He or she may be able to identify and remove endometrial implants, pelvic adhesions, or ovarian cysts, or to open blocked fallopian tubes. To do this, other instruments are inserted either through the laparoscope or through additional small incisions in the abdomen. These instruments may include *electrocautery* instruments (which remove diseased tissue through high-current electrical impulses) or lasers.

Possible Complications Laparoscopy usually has only minor risks and side effects. Complications are more likely, however, in patients who have had previous abdominal surgery, especially for pelvic or bowel adhesions. A previously undiagnosed case of PID also increases the risk of complications, as does obesity. Still, complications among young, healthy women occur in only about 3 out of every 1,000 cases. These complications include injury to structures inside the abdomen, such as the bowel, a blood vessel, or the bladder. When they do occur, surgery can be performed to correct the damage.

After you wake up, your abdomen may feel tender and bruised. You may have an ache in your shoulders from the gas that was used to distend your abdomen, and you may feel nauseated or weak from the general anesthesia. All of these side effects are normal and will fade within a few days.

Hysteroscopy As with laparoscopy, doctors use *hysteroscopy* both to diagnose problems and to treat abnormalities without the need for a second operation. Here, the inside of the uterus is viewed directly through an instrument inserted into the vagina and through the cervix. A sample of the endometrium (the uterine lining) can be collected for examination, and small growths such as polyps or fibroids can be removed.

Local anesthesia is usually used for diagnostic hysteroscopy when performed in the physician's office. Injections of anesthetic are made into the nerves in the uterus and into the tip of the cervix. The patient remains awake and aware at all times. The procedure can also be done in the hospital under local, general, or regional anesthesia.

The Procedure After the administration of anesthesia, the surgeon will dilate (widen) the opening in the cervix. As in laparoscopy, a gas is used to distend the area to be examined—this time it is introduced into the inside of the uterus to expand it. The doctor will grasp the edge of the cervix with a tenaculum and then will insert the hysteroscope—a long, slender instrument with an eyepiece and light at one end, somewhat like a laparoscope—through the cervix and into the uterine cavity (Figure 4.6). He or she can insert other instruments through the hysteroscope in order to collect a sample of endometrium or to remove growths inside the uterus. Sometimes laparoscopy is performed at the same time through an abdominal incision.

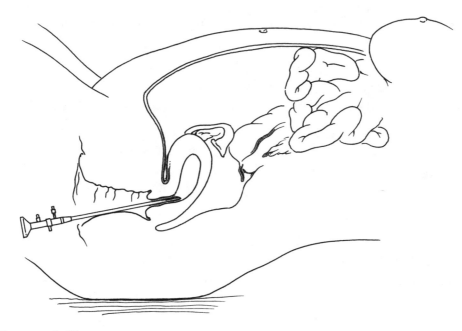

Figure 4.6: Hysteroscopy.

Possible Complications After the procedure, you can probably go home, unless there are complications or further surgery is needed. You may feel tender in your lower abdomen, and your shoulders may ache as the gas used to distend the uterus slowly dissipates. You may also feel nauseated and weak or have cramping or vaginal bleeding for a day or two afterward.

Making a Diagnosis: Possible Causes of Infertility

The various procedures used during the infertility workup are all performed with one object: to find a reason for a couple's infertility. If they can be diagnosed correctly, most of the causes of infertility can be successfully treated. This section presents information about a number of the possible reasons for a couple's infertility that may be uncovered during the infertility workup. Although these causes are discussed individually, keep in mind that for many couples—about 15 percent of cases—the cause is due to a combination of both male and female factors.

Female Infertility

Approximately 40 percent of infertility cases are found to be due to a female factor alone. These problems range from abnormalities in the anatomy of the reproductive organs or structures to hormonal problems that interfere with normal ovulation.

Ovarian Factors Problems with the ovaries may involve the ovaries themselves or their ability to produce viable eggs. Infection of the fallopian tubes can affect the ovaries, causing infertility. Hormone deficiencies can prevent ovulation.

Polycystic Ovarian Syndrome An ovarian cyst is a fluid-filled structure, somewhat like a blister, that forms on an ovary. Normally, an ovarian follicle grows throughout the menstrual cycle, eventually releasing an egg and then dissolving. When this process does not take place normally, the follicle may continue to enlarge and eventually form a cyst. In *polycystic ovarian syndrome* (PCO), multiple cysts are formed on one or both ovaries, creating a toughened outer layer and preventing ovulation (see Figure 4.7).

Researchers are not entirely certain of the cause of PCO, but the condition appears to be related to the production of reproductive hormones. Women whose thyroid or hypothalamus (the organ at the base of the brain that triggers the production of reproductive hormones) is not functioning properly may be prone to PCO. Obesity may also be related to PCO.

Symptoms of PCO include irregular or absent periods, acne, oily skin, abnormal hair growth such as facial hair, and pain in the abdomen during certain times in the menstrual cycle or during intercourse. The condition

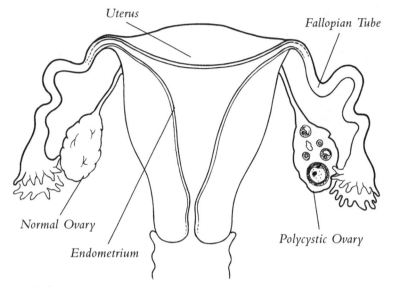

Uterus

Fallopian Tube

Normal Ovary

Endometrium

Polycystic Ovary

Figure 4.7: Polycystic ovary.

can be diagnosed through laparoscopy and is usually confirmed with ultrasound. When PCO is found in a woman with infertility, ovulation induction is usually used to stimulate the production of an egg (see Chapter 5, page 99 for more on this procedure).

Early Menopause At menopause, the body stops producing estrogen, and ovulation ceases. In most women, this natural process occurs between the ages of 42 and 56. About 1 percent of women under age 40, however, experience early menopause, also called *premature ovarian failure*. These women have the same symptoms as older menopausal women, including hot flashes, mood swings, and vaginal dryness. They also have the same long-term risks associated with normal menopause—heart disease and osteoporosis. These symptoms can be relieved and the risks lowered by taking estrogen replacement therapy. Although this treatment can also cause menstrual periods to resume, the ovaries will not produce eggs.

There are a number of possible causes of early menopause:

- **Genetic factors.** Some chromosomal disorders result in disruption of normal ovarian function.

- **Autoimmune disorders.** In autoimmune disorders, such as Addison's disease, myasthenia gravis, rheumatoid arthritis, systemic lupus erythematosus, and certain types of thyroid diseases, the body forms antibodies that attack the ovaries.

- **Damage to or destruction of the ovaries.** Radiation, chemotherapy, surgery, toxins, or certain infections of the reproductive tract can damage or destroy the ovaries or eggs.

- **Unknown factors.** In up to one-half of women who undergo premature ovarian failure, no clear cause can be identified.

Hormone tests for LH and FSH, performed as part of the infertility workup, may indicate early menopause. If FSH levels are elevated (as they are in women with premature ovarian failure), your specialist may watch for fluctuation over several weeks or months before a conclusive diagnosis is made. If you are diagnosed with early menopause, your doctor may perform additional tests to rule out thyroid, adrenal, or chromosomal abnormalities.

Women with premature ovarian failure rarely ovulate on their own, and spontaneous pregnancy would be rare. The use of so-called "fertility drugs" such as clomiphene or gonadotropins is rarely successful. Oocyte donation is the preferred treatment for infertility resulting from early menopause. (See Chapter 5 for details about these treatments.)

Tubal Factors Problems with the fallopian tubes are the cause of infertility in up to one-quarter of cases of female infertility. Tubal infertility is due to damage that results in partial or complete blockage of one or both tubes. This damage may occur anywhere along the length of the tube. Proximal tubal damage is located in the end of the tube near the uterus, and distal tubal damage is located in the end away from the uterus. As examples, Figure 4.8 shows distal tubal blockage from ovarian cysts (left side of illustration) and proximal tubal blockage from scarring inside the tube (right side).

Tubal infertility can take several forms:

- Damage to the *fimbriae*, the fingerlike projections at the end of the tube near the ovary, can prevent them from picking up an egg released from an ovary and directing it into the tube.

- Damage to the cells lining the tube may prevent or greatly reduce the chances of fertilization.

- Blockage in the tube can prevent sperm from reaching the egg.

- Partial or complete blockage of the tube can also prevent the fertilized egg from moving to the uterus, thus resulting in a tubal (ectopic), pregnancy (implantation of the fertilized egg inside the tube instead of in the uterus; see Chapter 2, page 26).

Tubal damage has several possible causes. Any distortion or constriction of a fallopian tube can prevent an egg and a sperm from meeting, or increase the likelihood of tubal pregnancy. In some cases, the damage can be reversed with surgery. When damage is extensive, a woman may still be able to carry a pregnancy to term using an ART (assisted reproductive techniques) method such as in vitro fertilization (discussed in Chapter 5).

Tubal Ligation Women who have undergone *tubal ligation* have had surgery to cut, clip, or tie their fallopian tubes, usually to prevent pregnancy. Sometimes a woman changes her mind about having children after having had a tubal ligation. In this case, surgery may be done to reverse the procedure. The success of this surgery depends greatly on how the original procedure was done. (See Chapter 5, page 107 for details on reversal of tubal ligation.)

Infection Pelvic inflammatory disease (PID) caused by chlamydia and gonorrhea (see Chapter 2, page 26) can cause inflammation and scarring inside

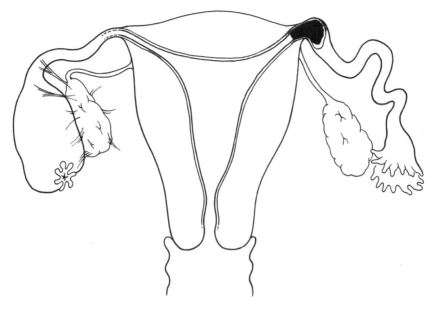

Figure 4.8: Distal tubal blockage (left) and proximal tubal blockage (right).

the fallopian tubes. Most often, PID results in proximal tubal damage, which can also be caused by *salpingitis isthmica nodosa* (SIN), a thickening and inflammation of the tubal wall.

Endometriosis Endometriosis is a condition in which tissue resembling the uterine lining forms outside the uterus. Endometriosis may bind the tubes and other structures together inside the pelvis, preventing ovulation or fertilization (see Figure 4.9).

Adhesions Adhesions consist of scar tissue that can form in the pelvis after surgery or as a result of infection or disease, binding together the surfaces of structures and organs inside the abdomen (see Figure 4.10). Surgery on the tubes or in other areas in the pelvis can result in tubal adhesions, especially if the surgery was extensive or involved a severe infection such as a ruptured appendix.

Cervical Factors Cervical factors may contribute to infertility, but they are rarely the sole cause. Infertility caused by a cervical factor has to do

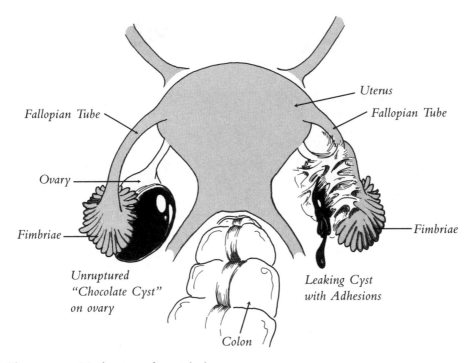

Uterus

Fallopian Tube

Fallopian Tube

Ovary

Fimbriae

Fimbriae

Unruptured
"Chocolate Cyst"
on ovary

Leaking Cyst
with Adhesions

Colon

Figure 4.9: Moderate endometriosis.

with characteristics of the cervical mucus through which sperm must travel in order to reach the egg. If mucus quality is poor or the quantity inadequate, the cervix may not be functioning properly. These causes of infertility are found through the PCT (described on page 70). The most likely explanation, however, is that the PCT was performed at the wrong time during the woman's menstrual cycle. Inadequate mucus production may be caused by prior surgery on the cervix.

A different cervical factor may also be the cause of recurrent miscarriage (loss of pregnancy before twenty weeks). In a condition called *incompetent cervix*, the cervix begins to dilate too early in pregnancy, causing the pregnancy to be lost. Most cases of incompetent cervix can be corrected through surgery, after which the woman can usually carry a pregnancy to term.

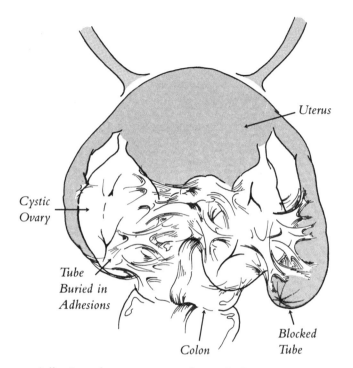

Uterus

Cystic Ovary

Tube Buried in Adhesions

Colon

Blocked Tube

Figure 4.10: Adhesions due to severe endometriosis.

Uterine Factors Abnormalities of the uterus can cause infertility by preventing implantation of a fertilized egg or by increasing the incidence of miscarriage. Uterine factors that cause infertility include the following:

- **Scar tissue** may be present inside the uterus from a pelvic infection, surgery, or disease.

- **Polyps** are clumps of endometrial tissue that protrude into the uterine cavity.

- **Uterine fibroids** are masses of muscle tissue that may form on the outside of the uterus, inside the uterine cavity, or within the uterine wall itself (see Figure 4.11).

- **An abnormally shaped uterine cavity** may be present in women with certain types of birth defects, such as those caused by exposure to DES (see pages 33 and 87).

- **Asherman's syndrome**, or **intrauterine adhesions**, is a condition in which fibrous tissue forms inside the uterine cavity. The space inside the uterus may be partly or totally filled as a result. Asherman's syndrome may develop after surgical procedures inside the uterus, such as *dilation and curettage* (D&C). This procedure, in which the contents of the uterus are removed by mechanical means, may be done to end a pregnancy or to remove tissue left in the uterus after a miscarriage. Asherman's syndrome can be diagnosed with HSG or with hysteroscopy.

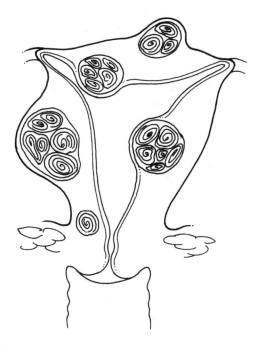

Figure 4.11: Fibroids.

✺ DES DAUGHTERS AND SONS

Diethylstilbestrol (DES) was prescribed for millions of pregnant women during the 1940s through the 1960s in efforts to prevent miscarriage. Unfortunately, it was later found that, not only was the drug ineffective for this purpose, it led to birth defects in the babies carried by these women. These babies came to be called "DES daughters" and "DES sons."

Women whose mothers took DES when they were pregnant with them are more likely to have changes in the cells lining the vagina and cervix, which may contribute to cervical-factor infertility. They also have a higher incidence of abnormally shaped uteruses and tubes, including a so-called "T-shaped" uterus. These abnormalities may make it more difficult for a fertilized egg to implant in the uterine lining. There may also be an increased risk of PID and tubal pregnancy in these women.

There has been much less research on DES sons than on DES daughters. Most of the effects of DES exposure on men have to do with abnormal development of the urogenital tract. Research that has been done so far does not seem to indicate that most DES sons have fertility rates that are any lower than nonexposed men. The exception is those DES sons in whom the drug has produced certain anatomical abnormalities. These include undescended testes, as well as hypospadias.

Male Infertility

A variety of factors can affect a man's fertility—infection, hormone imbalance, injury, or anatomical abnormalities. These or other problems in the male reproductive tract contribute to about 40 percent of all cases of infertility.

Semen Parameters Semen analysis can reveal problems with sperm production (the most common reason for male infertility), inadequate motility of sperm, or too many abnormal sperm in the semen sample. Parameters for normal semen analysis results are shown in Table 4.1. A normal sample should contain a total of at least forty million sperm, at least half of which are actively motile and normally shaped.

Oligospermia and Azoospermia Low numbers of sperm—usually defined as fewer than four million in the semen from one ejaculation—is called *oligospermia*, whereas the complete absence of sperm in the ejaculate is called *azoospermia*. These conditions are found in many infertile men. They may be caused by testicular disease, such as an infection that has destroyed the cells that make sperm. For example, when mumps is contracted after puberty, it sometimes lowers or obliterates a man's sperm count. Because testicular disease probably does not affect all the reproductive hormones, a man retains normal sexual function as long as his testosterone production is maintained.

Table 4.1 Normal Values of Semen Analysis

Volume (amount of semen ejaculated)	≥ 2 milliliters
Sperm concentration	≥ 20 million per milliliter of semen
Total number of sperm in ejaculate	≥ 40 million
Number of motile sperm	≥ 50 percent
Number of normally shaped sperm	≥ 30 percent

From McClure, R.D.: "Male Infertility." In: Keye, W.R., Chang, R.J., Rebar, R.W., Soules, M.R., *Infertility: Evaluation and Treatment.* Philadelphia, PA: W. B. Saunders, 1995, p. 66.

Hormone Deficiency Hormone deficiencies are another, but relatively rare, cause of low sperm output. Medications can increase levels of pituitary hormones. Unfortunately, for most men with a low sperm count or low motility, the cause is rarely hormonal, and the chance of success is, therefore, low with hormone therapy.

Anatomical Problems Abnormalities of the male reproductive tract are infrequent. Congenital abnormalities are conditions that are present at birth. Although there is little evidence that they are passed directly from fathers to sons, some are genetically acquired, such as hypospadias.

Varicocele About 15 percent of men in the United States have *varicoceles*— varicose veins above one or both testicles. Varicoceles pose no threat to a man's health and usually do not cause other serious problems, although they can be associated with testicular discomfort. Up to one-half of these men, however, have reduced fertility potential, possibly due to elevated temperatures within the testes that hinder sperm production. In some men, the varicose veins can be removed or tied off, which may improve their chances for initiating a pregnancy.

Obstruction An obstruction of the duct system within the male reproductive tract may block the outflow of sperm. Infection or surgery can scar the delicate tubules of the epididymis or obstruct the vas deferens. In other cases, the vas deferens has been severed or blocked off either deliberately, as in a vasectomy (male surgical sterilization to prevent pregnancy), or inadvertently during a hernia repair. In rare cases, the vas deferens is not present from birth.

If obstruction of the vas deferens is complete on both sides for any reason, the ejaculate will contain no sperm, even if the rest of the reproductive system is in working order. Surgical techniques can restore the free flow of sperm from the testes through the duct systems. The likelihood of success depends greatly on the cause of the obstruction, how long it has been present, and the microsurgical skills of the physician.

Hypospadias *Hypospadias* is an abnormal positioning of the opening in the penis. Instead of being located at the end of the penis, the opening is a lengthened slit on the underside of the penis. This can result in infertility if the delivery of sperm into the female reproductive tract is impaired, so that not enough of the sperm can find their way to the egg. Because this abnormality is sometimes related to the abnormal function of reproductive hormones, men with this condition may have other factors affecting their fertility as well.

Cryptorchidism In a condition called *cryptorchidism*, or undescended testicle, one or both testes have not descended into the scrotum. During male fetal development, the testes begin to develop inside the abdominal cavity. By about the seventh month of gestation, the testes have begun to move downward into the scrotum. In most baby boys, this process is complete by the time of birth. In the rest, the testes fully descend in the first year of life. In a small number of boys, however, one or both testes remain inside the groin or abdominal cavity. An operation can correct this condition, but fertility rates are decreased, even after surgery.

Infections Infections of the male reproductive tract can sometimes go unnoticed, but they nevertheless can affect fertility. In most cases, these infections are thought to develop from bacteria that have found their way through the urethra into the reproductive tract. Bacteria in the bloodstream (sepsis) can also spread to the genital tract.

Epididymitis Partly covering each testis is the epididymis, the glandlike structure where sperm are stored. *Epididymitis*, or infection of the epididymis, may develop from an untreated STD, such as chlamydia or gonorrhea, or from an infection of the urinary tract. In other cases, such infections may spread from an infection of the prostate gland. Men who have had surgery to remove the prostate may be at increased risk for epididymitis.

The first signs of epididymitis may be a low-grade fever, chills, and a feeling of heaviness in the testicles, which may be tender, sore, and swollen.

Severe infection may cause pain during urination or intercourse, or traces of blood in the semen. The infection can be diagnosed by examination. Antibiotics can be prescribed to treat the infection, and the patient may be advised to use cold packs to relieve the discomfort until symptoms have subsided.

This infection can impair fertility because the infection and accompanying increased body temperature can kill sperm stored in the epididymis. Timely and appropriate treatment usually restores fertility.

Prostatitis The prostate gland is composed of tiny sacs through which sperm pass on their way to the urethra. Bacteria from a nearby infection—such as in the epididymis, bladder, urethra, or colon—can find their way into these sacs and begin to grow there. *Prostatitis* is commonly caused by *Escherichia coli*, a bacterium found in the colon and large intestine. STDs, including chlamydia, gonorrhea, and trichomonas, can also lead to prostate infection.

There are three main types of prostatitis: acute, chronic, and nonbacterial. *Acute prostatitis* comes on quickly and produces noticeable symptoms, which may include fever and chills, pain in the lower abdomen, and burning during urination. If the prostate has swollen because of the infection, there may also be difficulty urinating. Like acute prostatitis, *nonbacterial prostatitis* also develops suddenly, but a culture reveals no bacteria in the urine. *Chronic prostatitis* develops gradually and continues for a prolonged period. The symptoms are very mild or may be absent, so that a man may not even realize that he has an infection.

The inflammation of the prostate that occurs with prostatitis can be lethal to sperm. Antibiotics are used to treat prostatitis. In the case of chronic infection, treatment may need to continue for as long as eight weeks, and recurrence is common. With persistent treatment, however, fertility can be restored.

Radiation and Chemotherapy Treatment of cancers of the reproductive tract, such as prostate cancer, often involve the use of radiation, which is toxic to sperm. The degree of damage to sperm production depends on how much radiation is used and for how long. With low levels of radiation, sperm production may take up to eighteen months to be restored. Higher doses can require up to five years for sperm production to return to normal, if at all.

Many drugs, such as chemotherapy (used to treat cancer), will also kill sperm. Others, such as cimetidine and spironolactone, can decrease male hormone production. Sulfasalazine and nitrofurantoin may affect sperm motility. Men who are about to undergo radiation and/or chemotherapy for cancer can arrange to have sperm frozen for later use to produce a pregnancy.

Occupational exposure to radiation, such as in men who work in nuclear power plants, is associated with changes in sperm cells that affect fertility. Proper shielding and limiting radiation exposure lessen the risks posed from this radiation source.

Ejaculation Disorders   Disorders of ejaculation take several forms and have a number of possible causes. Several steps must take place for ejaculation to occur, and a problem at any point in this process can interfere with normal ejaculation:

1. During emission, semen passes from the epididymis, into the vas deferens, through the seminal vesicle, and is deposited into the posterior urethra.

2. As semen passes through the prostate gland, muscles around the neck of the bladder tighten to close off the flow of urine.

3. Muscles controlling the opening of the urethra relax, allowing the semen to pass through. At the same time, the bulbourethral muscles contract, pushing the semen out through the urethra.

Ejaculation disorders may be due to anatomical defects in the urogenital tract, to problems with the nerves that control the muscles involved in ejaculation, or to psychological causes. Some of the main types of ejaculation disorders are:

- *Retrograde ejaculation*, in which semen backs up into the bladder instead of exiting out through the urethra. One possible cause is prior surgery on the bladder, prostate, or colon, which may have resulted in injury to the muscles involved in ejaculation. Other causes have to do with the part of the nervous system that controls these muscles.

- *Anejaculation*, or absence of ejaculation, is a condition in which no semen is expelled at all. This condition is caused by damage to the nerves that control the muscles involved in ejaculation. It is most common in men who have suffered injury to the spinal cord. Other causes may be neurological disorders, such as multiple sclerosis, diabetes, or certain types of surgery.

- *Delayed ejaculation* is a condition in which ejaculation occurs only after a very prolonged period. This condition is associated with certain types of antidepressant medications, neurological disorders, or psychogenic disturbances.

- *Premature ejaculation* is not an uncommon complaint in men under age 40. In this condition, ejaculation occurs before the penis is inserted into the vagina. The cause of premature ejaculation has always been thought to be psychological or related to sexual technique, but more recently, studies have suggested that a neurological factor may be involved in some cases.

Impotence Several medical conditions can cause or contribute to problems with achieving or maintaining an erection, a problem known as *impotence*. Like ejaculatory disorders, impotence may have medical, neurological, or psychological causes. Many cases of impotence are due to a combination of all of these factors.

During an erection, blood vessels in the penis relax, allowing blood to flow into them. Smooth (involuntary) muscles in the penis relax and compress small vessels through which blood drains, thereby trapping the blood and causing the penis to stiffen and become erect.

Medical conditions in which blood flow is hindered include diabetes, hypertension (high blood pressure), and atherosclerosis. Damage to the nerve pathways that supply the muscles and blood vessels in the penis is another possible cause. Such damage may be due to neurological disease, to spinal cord injury, or to lesions (abnormal growth) on these nerve pathways. Problems with the secretion of male hormones may also play a role. Finally, some causes of impotence are psychological in nature. Emotional stress, anxiety, depression, and relationship problems can all play a role in impotence.

A NEW TREATMENT FOR IMPOTENCE

In March 1998, the U.S. Food and Drug Administration approved a new drug, sildenafil (Viagra), for the treatment of impotence. The drug works by increasing the levels of certain chemicals that are involved in the relaxation of smooth muscle in the penis, allowing blood to flow more freely into the erectile tissue. Sildenafil does not cause erection in the absence of sexual stimulation.

In test subjects, sildenafil has been shown to increase the frequency, duration, and firmness of erections, and consequently to enhance sexual desire and satisfaction. The drug is effective even in patients with medical conditions that contribute to impotence, such as diabetes, hypertension, kidney disease, spinal cord injury, and depression.

Sildenafil is taken by mouth anywhere between thirty minutes to four hours before intercourse. It should not be taken more than once a day. The use of this drug may be of significant help to couples whose infertility has been due to impotence.

Sildenafil may cause potentially serious complications in some patients, especially those who have heart conditions or are taking certain medications. Men should receive a thorough consultation and examination from their doctors before taking this drug.

A Healthy Outlook

Diagnosing infertility is often a lengthy process fraught with anxiety and frustration. It may end without a specific diagnosis for your infertility. On the other hand, taking advantage of the tools available for diagnosing infertility may help you find a way to conceive and bear a child. Whatever the outcome, it is important to enter into the diagnostic process with open eyes and a realistic attitude. Through the tests that are used to diagnose infertility, you are finding out what it is possible to know, and your choices regarding treatment will become clearer once you have all the available information.

5

What's Next? Developing a Treatment Plan

Ideally, medical decisions about treatment for a health condition should be based on what is known about its causes. In most cases, this is also true for the treatment of infertility. By examining the treatment options described and explained in this chapter, and by considering them in light of the information gathered during the infertility workup, you and your fertility specialist can work out a plan to ensure the best chances of success.

This plan is likely to be a multistep process. Some treatments may require several attempts before you achieve pregnancy. Then, if a certain treatment proves unsuccessful, your specialist will advise you on the next step. The process can take a long time. However, infertility medicine is a growing field that promises new advances every year. Depending on the cause of your infertility, you may have very good chances of conceiving through one of the medical techniques described in this chapter.

Considering Your Options

Which treatments you and your specialist choose will depend on the conditions discovered or suspected during the infertility workup. These include lifestyle factors, as described in Chapter 2. For women, poor nutrition, extreme underweight, smoking, and other factors were discussed. For men, the overuse of hot tubs, tight clothing, and certain medications were discussed as possible reasons for a low sperm count. If factors such as these have been identified during the infertility workup, your first steps will be to change them.

If treatment is needed, you must consider your preferences and finances and discuss them with your specialist. There are many ways to treat infertility, including very advanced and cutting-edge techniques. All of these methods, however, may not be suitable for every couple. The price for some treatments, especially for assisted reproductive techniques, is high; there are sometimes risks involved; and the entire process can be emotionally intensive and time-consuming.

Before embarking on a lengthy and costly series of treatments, it is a good idea to consider the following questions:

- How strongly do you feel about having your own biological child?

- How much risk are you prepared to accept?

- What level of treatment can you handle financially?

- How much time will you be able to invest in the process?

This is a good point at which to revisit the section "Developing Your Decision-Making Skills" in Chapter 3 (page 39). The pointers given there may also come in handy at other stages of the treatment process.

The techniques described in this chapter are discussed in general order from least to most complicated and invasive. Knowing what your options are will help you work with your doctor in deciding which treatments make the most sense for you.

Nonsurgical Treatment

When the cause of infertility is believed to be an ovulatory disorder or cervical factor, simple and relatively noninvasive approaches to achieve fertilization are usually the first line of treatment. These approaches are also used for certain types of male infertility, including low sperm count, poor sperm motility and/or morphology, and disorders of erection or ejaculation. (See "Male Infertility" in Chapter 4 for information on nonsurgical treatments for men.)

Ovulation Prediction and Timed Intercourse

Choosing a method to predict ovulation allows a couple to time intercourse to coincide as closely as possible with ovulation (see the section "Tests for Ovulation" in Chapter 4, page 63). In selected cases, such as those in which a couple has been trying to become pregnant for a relatively short period and there are no immediately obvious causes of infertility, the couple may be advised to try timed intercourse before going on to more complicated procedures.

Ovulation Induction

Women with ovulation disorders can take medication that stimulates the ovaries to produce multiple mature follicles and ova. When the fertility workup reveals that a woman does not regularly ovulate on her own—usually because of conditions such as polycystic ovarian syndrome (PCO) and hormone disorders—*ovulation induction* is a noninvasive, nonsurgical treatment that may be an option. Several types of medications are available to treat ovulation disorders, depending on the nature of the problem.

With all of the methods of ovulation induction described here, ovulation may be detected in several ways. These include kits sold over the counter to measure levels of luteinizing hormone (LH), which peaks sharply at ovulation; measuring hormone levels in a blood sample taken in the clinic; and ultrasound to detect the presence of mature follicles.

All methods of ovulation induction increase the risk of pregnancies with twins, triplets, or even quadruplets and quintuplets. Multiple pregnancies typically have a higher risk of complications. (This topic is discussed further in Chapter 7.)

Clomiphene Women who do not ovulate regularly because of conditions such as PCO often use the drug *clomiphene citrate* (brand name, Clomid). It is also used to produce multiple follicles to maximize the success rates of certain types of assisted reproductive techniques. When used for this purpose, ovulation induction is referred to as *controlled ovarian hyperstimulation* (see "IUI with COH," page 105).

Clomiphene is taken by mouth during several days of the menstrual cycle (for example, days 3 to 7 or 5 to 9) to induce ovulation. Women who have very irregular periods may first use a medication such as *medroxyprogesterone* (a synthetic form of the hormone progesterone) to trigger the menstrual period.

Doctors usually start clomiphene at a dose of one fifty-milligram tablet once daily for several days. If this dose does not result in ovulation, it may be increased to two tablets a day for a few more days. If ovulation still has not occurred with this higher dose, the next step may be to increase the dose again, to three tablets per day, or to try another method of ovulation induction, such as injected gonadotropins. In some cases, your doctor may add another medication to increase the chances of ovulation while you are taking clomiphene. Examples of medications used for this purpose include *bromocriptine*, *dexamethasone*, and *metformin*.

Clomiphene is a simple and relatively inexpensive method of ovulation induction, but used alone it is less successful than other techniques. Pregnancy rates with clomiphene are much lower than ovulation rates— meaning that even if the drug is successful in inducing ovulation, pregnancy may not necessarily follow. The lower pregnancy rates with clomiphene alone are thought to be due to changes that the drug induces in the cervical mucus, in the lining of the uterus, and possibly in the eggs themselves.

Patients should also note that in a very small and possibly inconclusive study, clomiphene has been linked to ovarian cancer if taken for twelve consecutive months.

Gonadotropins Hormones called *gonadotropins* can be injected into a woman's muscle or under the skin to stimulate ovulation. These are either naturally derived or synthetic forms of follicle-stimulating hormone (FSH). FSH is produced naturally by the pituitary gland and triggers a follicle in an ovary to mature (see Chapter 2, page 17). Injected gonadotropins stimulate the ovaries to produce multiple follicles. Daily injections start early in the menstrual cycle and continue for up to two weeks. Ultrasound is used to determine whether any mature follicles are forming in the ovaries. When one or more mature follicles are seen, the woman is given an injection of another type of gonadotropin, called *human chorionic gonadotropin* (hCG). This triggers the release of the egg (ovulation) a day or two afterward.

This method of ovulation induction can be extremely successful and results in higher pregnancy rates than the oral clomiphene method. The main drawback is the significant time and discomfort involved in daily injections and frequent ultrasound and blood tests. Gonadotropins are also more expensive than clomiphene. Sometimes this type of treatment does not result in pregnancy on the first try and must be repeated over several more menstrual cycles, further raising the time and expense required.

As with all medications, the use of gonadotropins has possible side effects. One of these is a relatively rare condition called *ovarian hyperstimulation syndrome* (OHSS), which causes the ovaries to become enlarged and tender. OHSS occurs in only about 1 percent of cycles stimulated with ovulation-inducing drugs. It may occur late in an ovulatory cycle or early in a pregnancy conceived with the use of ovulation induction. In rare cases, hospitalization may be needed, but in most cases the problem resolves on its own in a short time. The risk of OHSS can be reduced by careful monitoring of follicle development throughout the woman's cycle.

Bromocriptine Excess production of the hormone prolactin (see Chapter 4, page 64) can be treated with bromocriptine, a medication that restores normal levels of this hormone. The medication is taken orally and can be very effective in restoring normal ovulation in women with mild to moderate elevations in prolactin production.

In some cases, high prolactin levels may indicate a growth, or tumor, on the pituitary gland. This is called a *pituitary adenoma*. The pituitary is a small gland located at the center of the brain. It secretes hormones that influence a woman's menstrual cycle. Most tumors on the pituitary gland are benign—meaning that they are not cancerous, grow slowly, and do not spread to other parts of the body. If a pituitary tumor is large enough, additional tests may be needed to decide how best to treat it.

The most common treatment is surgery, but depending on the size of the tumor, other approaches may be used. These include radiation to shrink the tumor, or oral medication to block the pituitary from producing excess hormones.

Recently, a new drug called *cabergoline* has become available to treat high prolactin levels, including those caused by pituitary adenomas. The advantages of this medication include the fact that it needs to be taken only twice a week. Its drawbacks, however, include its high cost.

Intrauterine Insemination

Insemination with sperm from the male partner or from a donor is used to treat infertility from a variety of causes. In *intrauterine insemination* (IUI), sperm are introduced directly into the woman's uterus, giving them a "head start" and allowing them to bypass the cervical mucus. For this reason, it is often used when the woman has cervical-factor infertility, when the man's sperm has lower-than-normal motility, or when poor sperm–cervical-mucus interaction is seen on the postcoital test (see Chapter 4, page 70). IUI may also be tried in couples for whom no clear cause of infertility can be found.

Because sperm must still find their own way into the fallopian tubes with IUI, this technique is not used if the woman has severe tubal damage or if the man has very poor sperm quality. Most centers also do not perform IUI in women who are past their early 40s, because the quality of a woman's eggs begins to decline at this time.

Before IUI is attempted, the fertility specialist will evaluate the fallopian tubes to ensure that they are open and will allow the passage of an egg. This is usually done with a hysterosalpingogram (HSG) or laparoscopy (both described in Chapter 4, pages 72 and 73).

When IUI is to be performed, semen is collected by masturbation in a sterile container. It is best to collect the specimen at the clinic in order to reduce the risk of contamination or environmental exposure, which can lower sperm quality. The semen sample then undergoes *sperm washing*, a treatment used to remove components other than sperm that are normally present in semen. These other components include hormones called prostaglandins and certain kinds of proteins.

During sperm washing, the semen sample is spun in a machine that uses centrifugation (rapid spinning) to separate the sperm from these other components in the semen. The entire procedure takes one to one-and-a-half hours and results in a product that is small in amount but contains many actively moving sperm.

For the IUI procedure, the woman is placed in a position like that for a Pap smear or pelvic exam, and a speculum is inserted into the vagina. A slender, flexible catheter is inserted through the cervix into the uterine cavity, and the semen is pushed out through the end. The semen may be deposited either in the cervix (see Figure 5.1, left side) or in the uterus (Figure 5.1, right side). The entire procedure takes only minutes.

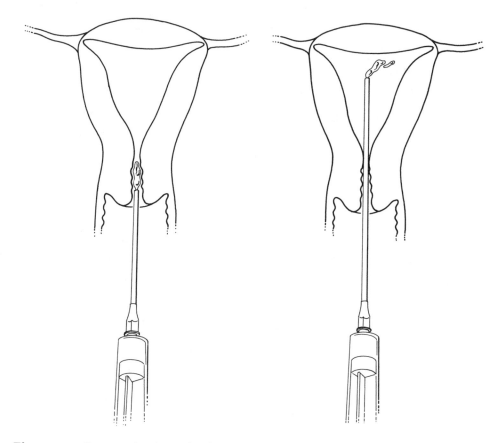

Figure 5.1: Intrauterine insemination.

The insemination procedure usually causes very little discomfort. Most women describe this procedure as feeling like having a Pap smear done. Some minor uterine cramping may occur during the procedure, but this does not affect the likelihood of pregnancy.

Iui can be done with sperm from either the woman's partner or from a donor. (Donor insemination is described later in this chapter, under "Third-Party Reproduction," page 117.)

Iui in Unstimulated Cycles When iui is performed without the use of ovulation-stimulating drugs (see "Iui with Coh," page 105), the procedure is timed to occur at the point in the woman's menstrual cycle when pregnancy is most likely to occur. Because sperm can live in the woman's reproductive tract for a few days, the procedure is usually done just before the time that she is expected to ovulate. The couple can continue to have intercourse at their usual frequency both before and after the procedure.

IUI in unstimulated cycles is a simple and relatively inexpensive procedure, but it results in pregnancy less frequently than other infertility treatments do. The procedure is usually repeated for a few cycles. If pregnancy does not occur, ovulation drugs may be tried.

IUI with COH So-called "fertility drugs"—such as clomiphene and gonadotropins—are often given before IUI to stimulate the ovaries to produce multiple eggs. This procedure, referred to as controlled ovarian hyperstimulation (COH), raises the chances of pregnancy by increasing the number of eggs available to be fertilized. Combined with sperm washing, COH allows multiple eggs to be exposed to an optimum number of healthy, motile sperm at the same time. This technique can be used to treat infertility due to absent or irregular ovulation or low sperm count.

COH is also used to develop multiple follicles during a cycle of in vitro fertilization. It is not effective for treating infertility caused by very poor sperm quality, premature ovarian failure, or very poor quality eggs due to aging.

Surgical Options

A number of surgical approaches can be used to correct problems that are causing or contributing to infertility. In women, surgery can repair damaged tubes or remove abnormal growths, such as polyps or fibroids, from the uterus. In men, anatomical abnormalities can often be corrected with surgery. Other surgical procedures can be used to reverse surgical sterilization in both women and men.

It is estimated that as many as 1.5 million men and women in the United States have had a surgical sterilization procedure to prevent pregnancy. Before they undergo sterilization, patients are counseled that they should consider the procedure permanent, and they are advised not to proceed unless they are certain that they desire no more children. But people's circumstances are subject to change. When people remarry after a divorce or the death of a spouse, sometimes they find they are ready for children again.

Some people who lose children through accident or illness regret their decision to sterilize. Or they may simply decide that they do want another child after all. Whatever the reason, some women and men decide to try to have their operation reversed.

Surgical Treatment in Women

Fertility specialists sometimes recommend surgery when the workup uncovers an irregularity in a woman's uterus and/or fallopian tubes. The two main types of surgery used in these cases—laparoscopy and hysteroscopy—actually combine the purposes of both diagnosis and treatment.

Laparoscopy and Hysteroscopy Both laparoscopy and hysteroscopy are far less invasive and have fewer complications than open abdominal surgery, or laparotomy. (Chapter 4 provides details on how laparoscopy and hysteroscopy procedures are performed and their possible complications.)

In laparoscopy, a thin, telescopelike instrument is inserted through a very small incision, usually made in the navel. This allows the doctor to view directly the organs and structures inside the lower abdomen. With laparoscopy, the doctor can diagnose scar tissue, endometriosis, polyps, and fibroids, as well as abnormalities of the fallopian tubes and ovaries. Often these problems can be treated surgically by inserting additional instruments, either through the laparoscope or through another small incision made in the lower abdomen.

Hysteroscopy is similar to laparoscopy in that it allows direct viewing of the inside of the body. The difference is that no incisions are made for hysteroscopy—rather, the hysteroscope is inserted through a woman's vagina and cervix and into the uterus. This procedure is used specifically to look at the inside of the uterus; it cannot be used to view the outside of the uterus or the fallopian tubes or ovaries.

Hysteroscopy is used to diagnose and treat abnormalities such as fibroids, scar tissue, and polyps. It can also identify a *uterine septum*—a thin partition inside the uterus that divides it into two cavities. Uterine septa can be surgically removed during the same hysteroscopic operation.

Tubal Ligation Reversal In women, sterilization, or tubal ligation, can be done using laparoscopy or mini-laparotomy. The latter procedure involves a larger incision than laparoscopy, but is not as invasive as open abdominal surgery. The fallopian tubes can be sealed either by tying the tubes or by using a clip, a band, or an electric current. The procedure is more easily reversed with the use of clips or bands than if an electric current was used.

The success rates of *tubal ligation reversal* (TLR) have improved with new microsurgical techniques. Depending on a number of factors—including the woman's age, the type of tubal ligation she had, and the lengths of the fallopian tubes left—50 to 80 percent of women having sterilization reversal will eventually become pregnant. However, the risk of ectopic pregnancy is slightly increased after TLR. Women considering this procedure should discuss success rates, possible complications, and risk factors with their doctors beforehand.

Surgical Treatment in Men

A number of causes of male infertility can be treated with surgery. Reversal of sterilization in men is being done more and more often, and success rates are improving. Surgery can also be used to correct anatomical problems in the male.

Vasectomy Reversal For men, surgical sterilization, or *vasectomy*, is a relatively straightforward procedure that is done on an outpatient basis. The patient usually receives local anesthesia, and a tiny incision is made on each side of the scrotum, or a single incision is made in the middle. A small segment of each vas deferens on each side is then removed. The two ends of the vas are sealed by the use of sutures, clips, or an electrical current.

Vasectomy reversal is the reconnection of the severed ends of the vas deferens. The procedure is accomplished with microsurgery, in which very fine needles and suturing material are used to rejoin the tubes.

The success rates of vasectomy reversal are much higher today than in the past years. Still, it is important to understand how success rates of vasectomy reversal are defined. Technically, a "success" is defined as the presence of sperm in the semen after vasectomy reversal. But the presence of sperm does not guarantee that the man will be able to father a child.

Corrective Surgery for Anatomical Problems Abnormalities of the male reproductive tract can sometimes be corrected with surgery. The success of surgical treatment for anatomical problems depends on a number of factors, including the severity of the defect. The most common anatomical problems that cause or contribute to infertility in men, and for which surgery can be performed, include the following:

- Varicocele

- Ejaculatory duct obstruction

- Hypospadias

- Undescended testicle (cryptorchidism)

Chapter 4 describes these problems and how they affect fertility in men.

Assisted Reproductive Techniques

The birth of Louise Brown, the first "test tube baby," was big news in 1978. She was the first child to have been conceived by an exciting new technique called in vitro fertilization, or IVF. Today, the procedure has been joined by many other *assisted reproductive techniques* (ART) that are available for infertile couples. Since the birth of Louise Brown, thousands of babies have been conceived and born each year with the help of these techniques.

ART may no longer be headline news, but it is still a source of debate and controversy. The ability to affect the process of human reproduction raises many ethical questions in the minds of many people, including scientists, doctors, researchers, and patients. These issues are explored in detail in Chapter 6.

Established forms of ART, such as IVF and its variations, are now widely accepted medical treatments for infertility. For many couples who have exhausted traditional clinical and surgical treatments, these techniques may offer the best hope for pregnancy.

In Vitro Fertilization

The Latin term *in vitro* translates literally as "in the glass." The term refers to procedures done outside of a living body, in a laboratory or other artificial environment. As its name implies, IVF techniques involve joining an egg and a sperm in a laboratory dish. If fertilization occurs, the resulting preembryo is then transferred to the inside of a woman's uterus, where, it is hoped, it will implant and grow into a living fetus.

When IVF techniques were first developed, they were used primarily for women whose fallopian tubes were blocked, damaged, or absent. Today, IVF is also used to treat infertility caused by endometriosis, certain types of male-factor infertility, and imbalances in the woman's immune system. It is also used in many cases of unexplained infertility.

IVF entails a number of steps that may vary slightly from clinic to clinic. Overall, the procedure involves

1. stimulating the woman's ovaries to produce multiple eggs during a specific time during her menstrual cycle (as described in "Ovulation Induction," page 99),

2. retrieving eggs from her ovaries,

3. adding sperm to the eggs in a laboratory dish, and

4. transferring the fertilized eggs into the woman's uterus.

Step 1: COH COH is used to stimulate the production of several eggs for an IVF cycle. (See "Ovulation Induction" earlier in this chapter for details on this procedure.) In most cases, a GnRH analog (a drug that is chemically similar to GnRH) is used to suppress all stimulation of the ovaries by the woman's own natural hormones. This allows the ovaries to respond more effectively to COH. IVF may also be performed without COH, although the success rates of the procedure in unstimulated cycles are lower because when only one egg is produced, the chances of fertilization are reduced. Ultrasounds are performed throughout COH to measure the size of ovarian follicles. Patients should also note that long-term GnRH analog administration can cause osteoporosis.

Step 2: Egg Retrieval After COH, the presence of multiple eggs is detected in the woman's ovaries through the use of vaginal ultrasound. This is the same procedure used to retrieve eggs from the ovaries. While the woman is under anesthesia, an ultrasound probe is inserted into the vagina. Once mature follicles are identified in the ovaries, the doctor guides a needle through the wall of the vagina and into the follicles (see Figure 5.2). The

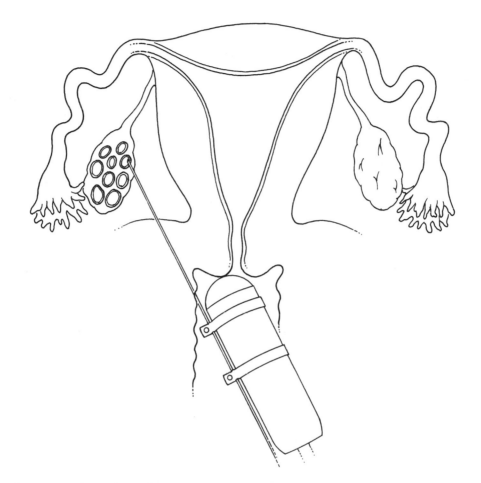

Figure 5.2: Egg retrieval.

eggs are then gently suctioned, or aspirated, out of the follicles and through the needle.

Egg retrieval can also be done by laparoscopy, although this approach is used less often than ultrasound. In this procedure, the eggs are removed through the laparoscope via a long, thin needle that is guided through the abdominal wall and into the mature follicles. General anesthesia is usually used for laparoscopic egg retrieval. (See Chapter 4 for details of how laparoscopy is performed.)

Step 3: Fertilization and Embryo Culture The eggs retrieved from the woman's ovaries are examined to determine their maturity. The stage of maturity of an egg will dictate when sperm will be added to it (insemination). Depending on the maturity of each egg, insemination may take place within minutes, hours, or days after retrieval.

Usually on the day of egg retrieval, the male partner produces a semen sample by masturbation. Sperm washing is carried out to obtain the most motile sperm. These sperm are then combined with each retrieved egg in a separate laboratory dish containing a growth medium. The dishes are placed in an incubator that maintains the same temperature as that of the woman's body.

Depending on the technique used (see "Micromanipulation and Microsurgery," page 114), it may take sixteen to eighteen hours for fertilization to take place. About twelve hours afterward, each embryo divides into two cells. The embryos may divide several more times while still in the incubator. Two to five days later, the fertilized eggs have each divided into multiple cells. At this point they are ready to be transferred into the woman's uterus.

Step 4: Embryo Transfer The next step, embryo transfer, is a relatively simple procedure that usually requires no anesthesia. One or more embryos suspended in a drop of culture medium are drawn into a transfer catheter, a long, slender tube with a syringe on one end. The tip of the catheter is guided through the cervix and deposits the embryos into the uterus (see Figure 5.3). The entire process takes only ten to twenty minutes and results in the transfer of one or more embryos. Although the transfer of multiple embryos increases the chances of multiple pregnancy (twins, triplets, or more), it also makes it more likely that at least one embryo will attach to the uterine lining and grow into a fetus.

Before the transfer procedure, a couple may decide with their doctor to freeze, or *cryopreserve*, additional embryos, which can be thawed and transferred at a later date. These embryos may be used if the first transfer procedure is unsuccessful or if the couple wishes to have more children.

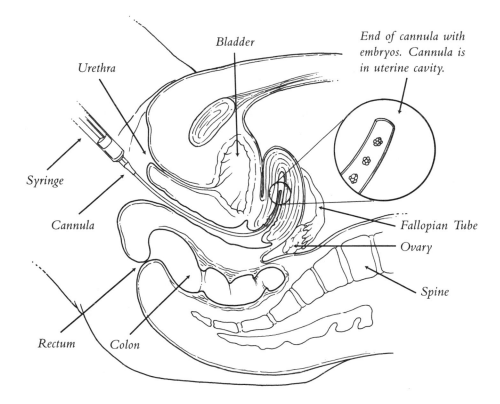

Figure 5.3: Embryo transfer. Patient is lying on her back with legs in stirrups. This is a side view.

Variations on IVF

Other types of ART resemble IVF, with variations on the procedure. These techniques include *gamete intrafallopian transfer* (GIFT), *zygote intrafallopian transfer* (ZIFT—sometimes called *pronuclear stage transfer*, or PROST), and *tubal embryo transfer* (TET). The chief difference among these procedures is the stage at which the eggs, sperm, and/or embryo are transferred to the woman's fallopian tube or uterus. All of these techniques involve stimulating and retrieving multiple eggs from the ovaries and preparing the semen sample as described for IVF.

Gamete Intrafallopian Transfer (GIFT) GIFT is the placement of unfertilized eggs and sperm together in the woman's fallopian tubes. Like IVF, a GIFT treatment cycle begins with COH, but the similarity ends there. With GIFT, eggs are retrieved using laparoscopy or vaginal ultrasound, and both sperm and eggs are placed in the fallopian tube during the same procedure. If the procedure is successful, fertilization occurs in the tubes as it does in natural, unassisted reproduction. Once fertilized, the embryo(s) travel to the uterus and implant there.

Zygote Intrafallopian Transfer (ZIFT) ZIFT differs from IVF primarily in the stage at which the fertilized egg is transferred. (This procedure is increasingly being referred to as pronuclear stage transfer, or PROST. Zygote is a term once used to describe an egg that has been fertilized but has not yet undergone cell division. The more common term for this stage of development is *pronucleus*.)

In ZIFT, eggs are retrieved by transvaginal ultrasound aspiration, as described for IVF, and are fertilized in a laboratory dish. The next day, before the fertilized eggs begin cell division, they are transferred into the woman's fallopian tubes by laparoscopy.

Tubal Embryo Transfer (TET) TET involves the transfer of a more developed embryo than that used in GIFT or PROST. With TET, a fertilized egg that has reached the four- to eight-cell stage of division is transferred into the fallopian tube. Transfer usually takes place about twenty-four hours after fertilization. The developing embryo then proceeds to move into the uterus as in an unassisted pregnancy.

Micromanipulation and Microsurgery

In cases where very few sperm are available or a severe male factor is causing infertility, the chances of fertilization can be increased by the use of micromanipulation or microsurgical techniques. Some of these techniques are designed to assist fertilization or implantation. Others are used to retrieve sperm in cases of obstruction of the male genital tract. All may be used in conjunction with IVF or other forms of ART.

Techniques to Enhance Fertilization or Implantation These techniques involve the use of microsurgical instruments to hold the egg while microscopic needles are used to inject sperm:

- **Intracytoplasmic sperm injection (ICSI):** A microsurgical needle is used to inject a single sperm directly into the egg to achieve fertilization.

- **Microinsemination:** Sperm are concentrated into a small drop of fluid and then placed around the eggs to increase the chances of fertilization.

- **Assisted hatching:** A micromanipulation technique performed after fertilization with IVF, assisted hatching is designed to improve the implantation of the embryo in the uterine wall. In this procedure, a microscopic hole is made in the zona pellucida to facilitate the release of the embryo from the egg membrane. The hole may be made with either a microsurgical needle or with chemicals that create a thin spot in the zona.

These micromanipulation techniques are relatively new procedures that may not be available in all fertility clinics. Their success depends on several factors, including the quality of the man's sperm, the age of the woman, and the experience of the clinic.

Techniques to Retrieve Sperm The advent of ICSI has greatly reduced the number of sperm that need to be collected in order to achieve fertilization. This advance has given rise to new techniques that hold promise for the treatment of some types of male infertility.

- **Testicular sperm extraction (TESE):** A procedure in which immature sperm cells are collected directly from the testes. Though not fully mature, these sperm cells still contain genetic material. Scientists are currently trying to find ways to combine this genetic material with that of mature ova in order to achieve pregnancy.

- **Round spermatid nuclear injection (ROSNI):** A form of TESE followed by ICSI. The nucleus of an immature sperm cell (a round spermatid) is isolated by TESE and then injected into an egg with ICSI. A few pregnancies have been achieved in humans with this technique, but at this time ROSNI remains on the cutting edge of new reproductive techniques. The procedure may hold promise for the treatment of azoospermia (in which no sperm are present in a man's semen) that cannot be corrected with surgery.

- **Microsurgical epididymal sperm aspiration (MESA):** A technique of collecting sperm directly from the epididymis (the organ where sperm cells mature and are stored). This procedure is an option for men with obstruction of the vas deferens that cannot be corrected with surgery. It is performed by making a tiny opening through the scrotum and into the epididymis. This incision is usually made with the aid of a microscope to reduce the risk of scarring the tissue inside the epididymis. Fluid from the epididymis is then collected and examined under a microscope for live, moving sperm. A number of sites within the epididymis may be checked in this manner, until viable sperm are seen. The sperm collected through MESA can then be prepared for insemination in an IVF, GIFT, or ZIFT procedure.

- **Electroejaculation:** A method of sperm collection in men with ejaculatory disorders (see Chapter 4, page 92). In this technique, a probe is inserted into the rectum and passes a low electrical voltage into the nerves that stimulate ejaculation. Semen is collected from the penis and examined for active sperm as described above.

Third-Party Reproduction

Under certain circumstances, a couple may wish to consider using *third-party reproduction* in order to become parents. The term refers to the use of sperm, eggs, embryos, or the uterus of a third person (donor), who may be either known to the couple or anonymous.

Third-party reproduction is attended by many complex social, ethical, and legal issues. To date there has been limited experience with this process, except in the area of sperm donation. As a result, areas of uncertainty exist about the long-term effects of these arrangements on the people involved.

Anyone who is thinking about third-party reproduction should get the facts about what is involved. They should also consult an attorney who is familiar with the legal issues surrounding the process. All parties agreeing to third-party reproduction should seek in-depth counseling concerning their legal rights and responsibilities, as well as their personal feelings about the arrangement. (The legal, ethical, and emotional ramifications of these and other aspects of ART are explored in Chapter 6.)

Therapeutic Donor Insemination (TDI)

Couples in whom male-factor infertility is not treatable may choose to have IUI with sperm from an anonymous donor. Known as *therapeutic donor insemination* (TDI), this technique can result in pregnancy with a baby who is biologically related to the mother but not to her partner.

Couples might choose TDI if the man does not produce enough healthy sperm due to, for example, a vasectomy that cannot be reversed, or because he has undergone chemotherapy or radiation treatment. Others who may choose this option include couples who know that one or both partners are carriers of a known hereditary or genetic disorder, such as Tay-Sachs disease, Huntington's disease, hemophilia, or chromosomal abnormalities. Single women and lesbian couples who want children sometimes also choose this option.

The procedure used for TDI is the same as that for any other type of IUI. The only difference is that the semen used has been collected from a donor, who may or may not be known to the couple.

Donor semen is always frozen and stored until it can be determined that the donor is negative for human immunodeficiency virus (HIV). Because it may take months for the virus to show up on the donor's blood test, all donor semen is isolated for at least six months before it is used for IUI. The donor is tested for HIV at the time of semen donation and then again six months later. If this second test is still negative, the donor semen is considered safe to use.

Anonymous donors provide semen to sperm banks. Couples who are considering using an anonymous donor from a sperm bank should ensure that the facility follows the standards of the American Society for Reproductive Medicine (ASRM). The sperm bank should provide a thorough medical history of the donor and his family. The donor should be under age 40, and it is preferable that he is known to have fathered a child in the past. The sperm bank should also test all donors for common diseases, Rh factor, hepatitis, sexually transmitted diseases, and HIV.

Some sperm banks provide information about donors' physical traits for couples wishing to have a child who resembles their family. Some also provide detailed information on personal habits, education, profession, hobbies, and talents. Although this information may be interesting, there is no assurance that habits and behaviors will be passed on to the child.

To avoid future legal problems, individuals using TDI should make sure that the sperm bank or the physician keeps a permanent, confidential record of the donor's health and screening information.

Sometimes individuals wish to use donated semen from someone known to them, such as a relative or a close friend. This practice can be fraught with problems, however, if the relationship with the donor changes. It can also be more difficult to keep secret the identity of the child's biological father, if that is what the couple wishes.

Donor Oocytes

Some women are poor candidates for assisted reproductive techniques such as IVF, GIFT, and ZIFT because they do not produce enough healthy eggs for insemination. These women may wish to consider using eggs from a donor, who may be either known to them or anonymous. When donor oocytes (the term used to refer to eggs before ovulation) are used, eggs are retrieved from the donor, fertilized in the lab, and then transferred to the uterus of the recipient, who carries the pregnancy to term.

Donor oocytes are not used for women with abnormalities of the uterus that would prevent them from carrying a pregnancy to term. This option may be a good choice, however, for women with a normal uterus whose ovarian function is declining or lacking due to factors such as:

- Ovaries absent at birth

- Advanced age

- Premature menopause

- Surgical removal of ovaries

- Chemotherapy or radiation therapy for cancer

FERTILITY OPTIONS AND CANCER TREATMENT

Some 26,000 reproductive-age adults who have been diagnosed with cancer are living in the United States today. For many of these people, a primary concern is how their reproductive potential will be affected by the cancer itself as well as its treatment. Advances in fertility medicine can offer the hope of having normal, healthy children to many of these men and women.

Cancer in an advanced stage is often accompanied by malnutrition and other problems that impair fertility. Treatment for cancer often involves the use of powerful drugs and radiation that can impair or damage reproductive function. In women, cancer of the ovaries can destroy ovarian tissue and lead to ovarian failure. Certain types of gynecological cancer often result in amenorrhea. In both sexes, cancer drugs destroy rapidly dividing cells, including those in the ovaries and testes. Women taking these drugs typically stop menstruating, and men often have poor semen quality.

Refinements in cryopreservation techniques may allow many cancer patients to preserve reproductive cells for future use, when their cancer is in remission and they are no longer receiving the harmful effects of cancer therapy. Cryopreservation techniques have been used for many years to preserve semen. More recently, it has been possible to cryopreserve embryos resulting from IVF. Although no satisfactory method for freezing oocytes has yet been developed, this may be a possibility in the not-too-distant future. Experiments are already underway to cryopreserve sections of a woman's ovaries for later use. Although this technique has not been shown to work in humans, there has been limited success in experimental animals.

These techniques can make it possible for some cancer patients to preserve their reproductive function until the harmful effects of cancer and its treatment are past. For example, a woman diagnosed with cancer can have her eggs fertilized with her partner's sperm and the embryos frozen until her cancer is in remission. A man can have his sperm frozen, or sperm can be collected and his partner inseminated, before he begins cancer treatment. Pregnancy rates are lower when frozen sperm or embryos are used, but there seems to be no increase in birth defects in pregnancies using frozen sperm or embryos.

A downside of embryo cryopreservation is that it requires a full IVF cycle, which may delay cancer treatment by four to eight weeks. Depending on the type of cancer and how advanced it is, it may be risky to wait this long to begin treatment. Patients contemplating cryopreservation to preserve fertility need to be fully informed of all the facts, including the likelihood of success and risks of the procedure.

Some couples may want to consider using donor oocytes if the woman carries a genetic defect that they wish to prevent passing to their children. This may also be a good option for women who have had several unsuccessful IVF cycles that were thought to be caused by poor egg quality.

Children born through the use of donor oocytes are not biologically related to the woman who carried the pregnancy and gave birth. For some couples, this is less important than for others, who may wish to have their own biologically related children. Some couples choose to use eggs from a close relative, such as a sister of the woman. Others prefer that the donor be anonymous and that her identity—and even the fact that donor eggs were used at all—be kept from the child and family members, at least until the child reaches the age of consent.

Evaluation of the Recipient Couple Before the use of donor oocytes is considered, a couple must undergo several procedures and tests:

- Thorough medical history of both partners

- Assessment of ovarian function in the woman (if her ovaries are present and she is not menopausal)

- Lab tests for both partners:

  - Hiv

  - Hepatitis B and C

  - Other sexually transmitted diseases (STDs)

  - If the woman is over age 40, possibly cardiovascular screening and a consultation with an obstetrician to identify any other factors that would make a pregnancy especially high risk

  - Possibly hysteroscopy or hysterosalpingography to evaluate the woman's uterus

If no reasons are found for the woman not to carry a pregnancy to term, the couple may proceed to find a suitable donor.

Screening and Recruitment of Donors   Donor oocytes can be obtained in one of three ways:

1. **Known donor:** The recipient or couple may know the egg donor, who may be a close relative or friend who is willing to donate her eggs.

2. **Anonymous donor:** The recipient or couple may not know the identity of the egg donor. This option is usually arranged through established egg donation programs.

3. **IVF programs:** Women in IVF programs may agree to donate excess eggs developed during their IVF cycles to other infertile patients.

All egg donors, whether known or anonymous, should be screened according to the standards set forth by the American Society for Reproductive Medicine. These standards specify that donors should be younger than age 34 (except in special circumstances, such as when a relative is to be the donor). Women in this age group are preferred because they respond more favorably to the hormone medications used for the procedure, and their eggs generally result in higher pregnancy rates. Natural (unassisted) pregnancies in women over age 34 are more likely to result in chromosomal abnormalities such as Down's syndrome. If the donor is over age 34, the doctor may suggest prenatal testing to detect certain genetic disorders if a pregnancy is established.

It is preferable for the donor to have had a prior pregnancy, whether or not it was carried to term. She should also be screened for genetic disorders, hepatitis B and C, HIV, other STDs, and Rh compatibility.

HIV is of special concern when donor eggs are used. Although HIV testing is extremely accurate, very early-stage infections, such as those that have been present for less than six months, can escape detection. Because the freezing process can damage the DNA inside eggs, cryopreservation of oocytes is still a highly experimental procedure. However, people concerned about the risk of HIV may want to consider having the donated eggs fertilized and the resulting embryos frozen for future use. The donor can then be retested after six months, and if she is still HIV negative, the embryos can be thawed and transferred to the recipient's uterus. The main drawback of this arrangement is that pregnancy rates with frozen embryos are much lower than with fresh ones.

The Procedure Assisted reproduction with donor eggs requires both the donor and the recipient to undergo certain procedures and monitoring. For her part, the donor must produce enough healthy eggs to be retrieved. The recipient must have her menstrual cycle adjusted with hormones to match the donor's. The timing of the entire procedure must be just right so that by the time the eggs are retrieved and fertilized, the recipient's uterus is in the right stage to receive the embryos.

Donor Preparation When donor oocytes are used, COH is used in the donor to stimulate the development of multiple eggs. A combination of hormone medications is generally given for two to three weeks before the egg retrieval. As the follicles mature, their development is monitored by ultrasound and measurement of blood hormone levels.

When the follicles are mature, ovulation is triggered by an injection of human chorionic gonadotropin (hCG). About thirty-four hours later, the eggs are collected by aspiration, graded for maturity, prepared for fertilization, and inseminated in the same ways as for any type of IVF procedure.

Recipient Preparation While the donor is undergoing COH, the recipient will have her menstrual cycle synchronized with the donor's. This is done by giving the recipient GnRH analogs to suppress her own menstrual cycle and hormones to stimulate the uterine lining to develop and thicken in preparation for pregnancy. The uterus must be ready to receive the embryos within a few days after the eggs are retrieved from the donor and fertilized in the lab.

Surrogate Gestation

Surrogacy is an arrangement in which a woman (the surrogate) carries a pregnancy for an infertile person or couple. The first publicly acknowledged surrogate pregnancy occurred in the United States in 1985. Since that time, couples have used the arrangement with increasing frequency, but it still is relatively uncommon, probably because of the potential for legal, ethical, and emotional problems.

Surrogate gestation may be arranged in one of two ways. In the first, the surrogate carries a pregnancy that was created from the sperm and egg of two other individuals. This arrangement involves the use of IVF, which is carried out just as it is when the female partner is to carry the pregnancy. The only difference in a surrogate arrangement is that the fertilized egg is transferred into the uterus of the surrogate rather than the female partner. In this type of surrogacy, the child that is eventually born is not biologically related to the woman who gives birth.

In the second type of surrogacy, IUI is used to impregnate the surrogate with sperm from the male partner. This results in a pregnancy with a child who is genetically related to the father, but not to his female partner. This type of arrangement, sometimes called *traditional surrogacy*, is the most common type of surrogacy. In these cases, the female partner or couple must legally adopt the child after birth.

Indications Surrogacy may be considered when the female partner has functioning ovaries but no uterus. For example, a woman may have had a hysterectomy or may have been born without a uterus. Surrogacy may also be considered when the woman has a malformed uterus that cannot be corrected, is unable to carry a pregnancy for other reasons, or has a medical condition that would make a pregnancy life-threatening.

Traditional surrogacy might be considered if a woman has no ovaries or does not produce healthy eggs because of declining ovarian function or premature ovarian failure. It might also be an option for women with genetic disorders that could be passed on to her offspring.

Screening and Evaluation of Surrogates A surrogate may be a close friend or relative of the couple, a woman who is paid for the service, or an anonymous individual whose confidentiality is protected throughout the entire process. Guidelines for surrogates are not as well established as those for sperm and egg donors. Ideally, the surrogate is a young, healthy woman who has carried a previous pregnancy to term without complications. She should have no risk factors that could endanger a pregnancy, such as a history of alcohol abuse, tobacco use, or other drug use. She should not have a history of engaging in high-risk sexual behavior or of medical disorders, such as diabetes or Rh sensitization.

A potential surrogate should undergo a complete medical history and physical exam, in addition to a screening for infectious diseases. She may also have her uterus evaluated. In addition, ASRM strongly recommends a psychological evaluation.

The Procedure Like the use of donor eggs, surrogate gestation involves the use of IVF. If the embryo is created from the couple's sperm and egg, the surrogate may be given hormones to prepare her uterus to receive the embryo. IVF is carried out as described earlier in this chapter, and the embryo is transferred into the surrogate's uterus.

Donor Embryos

Embryo donation involves donor eggs that have been fertilized with sperm from the donor's partner or from another male donor. Donor embryos may be used under the same circumstances as when donor eggs are used.

Embryos may be donated by a woman having an IVF procedure who becomes pregnant and does not wish to keep the fertilized eggs that were not used. These embryos may then be transferred to the recipient's uterus. A child resulting from a donated embryo will not be genetically related either to the woman who carries the pregnancy or to her partner.

As with egg donation, the recipient's uterus will be prepared to receive the donated embryo. Depending on the timing of fertilization and the recipient's menstrual cycle, donated embryos may be either fresh or cryopreserved (frozen). The success rate for embryos that have been previously cryopreserved is lower than for fresh embryos.

Like every other case of infertility, the specifics of your own particular case will be unique. Because treatment decisions need to be based on these specifics, no two cases of infertility will necessarily be managed exactly alike, even if they share some of the same causes and contributing factors. Still, some generalizations can be made about which treatments are best suited to which types of infertility:

- **Inoperable tubal damage:** IVF is often appropriate for women with blocked fallopian tubes from surgical sterilization that cannot be reversed, from severe endometriosis, or from scarring and adhesions that have resulted from tubal infection. Many IVF programs recommend the removal of severely damaged tubes before IVF is performed, in order to increase the chances of pregnancy and decrease the risks of infection after the procedure.

- **Unexplained infertility:** Ovulation induction with IUI, IVF, GIFT, or donor oocytes may be suitable for couples who remain infertile despite normal fertility evaluations.

- **Ovulatory disorders:** Donor oocytes are appropriate for women with premature ovarian failure and for any woman past the age of 43, even if she still has regular periods.

- **Male infertility:** Ovulation induction with IUI, GIFT, or IVF, with or without ICSI, may be used if treatment of male infertility is unsuccessful. Recently, TDI has become a more distant second choice for many couples because of increased success rates of these first-line treatments.

An Array of Possible Options—and Outcomes

The advances made in the treatment of infertility are staggering compared with just a few decades ago. The options available today have created new hope for parenthood among millions of men and women.

With these expanded options comes an increased need for education and understanding of what might—and might not—be possible for infertile couples. Despite the amazing array of new techniques to help these couples achieve pregnancy, there are still cases of infertility for which no cause can be found and no treatment seems to work.

Infertility treatment is a long and sometimes frustrating road for many couples. When successful, the effort is rewarded when a child is born. When unsuccessful, a couple may have to come to terms with childlessness, or accept alternatives to carrying a pregnancy or having their own biological child.

The complexities of infertility treatment encompass not just medical but also emotional, psychological, ethical, and legal concerns. Exploring these issues before embarking on a lengthy, costly, and sometimes frustrating course of treatment is wise for your own well-being and will help you to maintain a realistic attitude while exploring your options.

6

Legal, Ethical, and Emotional Issues Surrounding Infertility Treatment

Breakthroughs in the treatment of infertility are occurring at a faster rate than ever before. Accompanying these advances are a host of legal and ethical questions, as well as emotional issues for the couples involved in infertility treatment.

As more and more couples avail themselves of these new technologies, questions arise concerning increasingly complex issues—for example, the legal rights of women acting as surrogates, or how to resolve disputes about the custody of preserved embryos. Once a simple matter, the issue of who are a child's rightful parents is harder to decide when three or more people have had a hand in bringing the child into the world.

It is easy to see why the subject of infertility treatment is attended by intense emotions. For example, sperm or ova donated by a relative or an anonymous donor allow a woman to carry a baby that is not biologically related to her or to her husband. This technology can create complex feelings for each of the partners and can severely stress a relationship. Assisted reproductive techniques can bring the joy of a baby into a family who would otherwise remain childless, but unless the potential for emotional stress is anticipated and addressed before a couple begins, ART can create confusion and distress for all the involved parties.

This chapter explores some of the legal and ethical issues that arise surrounding infertility treatment. It also explores the many and varied feelings that are common to women and men undergoing infertility treatment. Suggestions are offered for improving communication between partners. Finally, a list of resources is included for further information and help.

Bioethics and the Law

As new ground is broken in the choices available to infertile couples, new laws are being forged to answer the legal questions that arise. Advances in reproductive technology, however, continue to proceed at a far faster pace than laws can be passed to guide decisions.

Laws are established to maintain order in society and to uphold justice in cases of conflict. Like laws governing other areas of life, those having to do with ART arise out of ethical concerns. Much has been said and written in the news media in recent years about the field of *bioethics*—the study of moral issues in medical treatment and research. Although this term may be relatively new, the concepts behind it are not. The field of bioethics is rooted in a long tradition of establishing ground rules for the practice of medicine.

Whereas scientific developments take place as rapidly as experts can make new discoveries, laws must be carefully considered before being put into practice. This time lag is the chief reason why, as one scholar has put it, the law is "marching with medicine but in the rear and limping a little."[1] It is also why there are few laws, at either the federal or the state level, that have much to say about the use of ART. The laws that do concern this area of medicine mainly have to do with two areas: the use of fetal tissue for research and experimentation, and questions of custody of children born through surrogacy arrangements.

A SHORT HISTORY OF MEDICAL ETHICS

Medical ethics has its roots in ancient writings such as the Hippocratic Oath. The American Medical Association was the first organization to establish, in 1846, a professional code of ethics. It was based largely on a treatise written by the English physician Thomas Percival in the eighteenth century. The Nuremburg Code governing the ethics of scientific and medical research on humans was established at the close of World War II in response to the outrageous and inhumane practices forced upon prisoners of concentration camps in Nazi Germany. Advances in reproductive medicine, from the 1950s to the present day, continue to present new challenges and questions that further complicate the moral and ethical issues of medical research and practice.

Laws governing medical practice represent attempts to regulate human behavior in response to ethical concerns. These laws arise out of ethical concerns that patients be treated humanely and with full recognition of their rights to fully understand and consent to medical tests and procedures.

---

[1] Sir Victor Windeyer, *Mt Isa Mines v Pusey* 45 ALJR 88 at 92, 1971.

The Law and Fetal Research

The first of the modern techniques to assist infertile couples was in vitro fertilization (IVF). Research into IVF and other methods to treat infertility increased dramatically in the years following World War II. The first "test tube baby," Louise Brown, was born in 1978. In the decades since, IVF and its variations have become widely accepted and practiced.

Like most scientific breakthroughs, IVF became widely accepted much sooner than laws could be established to regulate its use. Federal regulations prohibit the use of fetal tissue for research and experimentation, but these laws do not pertain to IVF because the law defines a "fetus" as "any product of conception from implantation to birth." IVF involves handling embryos before their implantation in a woman's uterus. Further, these laws were passed with abortion in mind, not in vitro techniques.

In addition to federal regulations, many states have passed individual laws that pertain to IVF. Some have adopted their own definition of "fetus," such as "a product of conception from the moment of fertilization," but many do not define the terms *fetus* or embryo. Some states specifically address IVF in their statutes governing fetal research, either including or exempting the procedure from its prohibitions on fetal experimentation.

It is common for more embryos to be produced during an IVF cycle than a woman can safely carry in a single pregnancy. Questions arise about what may be done with these embryos—whether they may be preserved, donated, or sold to another person or couple, or discarded. The answers to these questions depend on the laws of the state and the wishes of the couple, but sometimes such situations become complicated. Cases such as that of Mary Sue and Junior Davis (see below) illustrate how complex these issues can become—and how the law follows rather than anticipates medical advances.

✒ THE DAVIS CASE

An important issue raised by IVF is the right of a couple to decide what may be done with extra embryos produced during an IVF cycle. This question was addressed in a landmark case in Tennessee in 1989.

A married couple, Mary Sue and Junior Davis, received IVF treatment, but shortly afterward began divorce proceedings. At the time the Davises filed for divorce, seven embryos had been preserved for future implantation in Mary Sue's uterus. During the divorce, the couple disagreed about what should be done with the embryos. Mary Sue wished to preserve them, but Junior took the position that, should his wife decide to have them implanted, he should not be forced to bear financial responsibility for children that he did not agree to have.

The court awarded custody of the embryos to Mary Sue, but Junior appealed the decision. The appeals court then awarded custody equally to both parties but ruled that, just as Mary Sue could not be forced to implant the embryos against her will, neither should Junior be forced "to bear the psychological, if not the legal, consequences of [fatherhood] against his will." The embryos were eventually returned to Junior Davis and allowed to expire.

Couples who undergo IVF and who consent to cryopreservation of excess embryos usually state in writing their wishes concerning what may be done with the preserved embryos. In some cases, however, couples have not stated their wishes and cannot be contacted to make their wishes known. This poses a problem for the fertility clinic, which must then decide whether to continue storage of the embryos.

The American Society for Reproductive Medicine (ASRM) has taken the formal position that infertility programs should require each couple considering embryo storage to give written instructions about what is to be done with the embryos. This statement should specify the course of action to be taken in the case of the couple's death, divorce, separation, failure to pay storage fees, inability to agree on what is to be done, or lack of contact with the program. It should also state that the program is allowed to dispose of the embryos after a certain period if the couple has not kept the program informed of where to reach them. The couple can agree together to change the statement at a later time by signing a new statement that specifies their wishes.

Most fertility clinics rely on written statements such as this to guide them in decisions concerning stored embryos. Currently the law does not give clear guidance on what may be done with frozen embryos when a couple disagrees or cannot be contacted. For this reason, when there is no written statement, a couple cannot be sure of what will be done with their embryos under such circumstances.

The Law and Surrogacy

Surrogacy is much less commonly used than IVF, but nevertheless it has posed significant legal dilemmas. So far, these problems have usually centered on determining the rightful parents of a child born through a surrogacy arrangement.

Surrogacy goes to the very heart of the question, "What is a parent?" From a biological standpoint, a "mother" can be defined as a woman who conceived the fetus, carried the pregnancy, and gave birth to the child. In a social sense, the term "mother" refers to the woman who raises the child. Sometimes the biological and the social roles of motherhood are filled by two different women, as in the case of adoption.

Surrogacy breaks down the role of the biological mother even further by making it possible for as many as three different women to be involved in providing the egg, carrying the pregnancy, and raising the child. In such instances, the child could be said to have three mothers:

1. A *biological* mother, to whom the child is genetically related

2. A *gestational* mother, who carried the pregnancy and gave birth to the baby

3. An *intended* mother, who raises and cares for the child until adulthood

The situation can be further complicated if donated sperm is used, which makes it possible for the child to also have two fathers:

1. A *biological* father, to whom the child is related genetically

2. An *intended* father, who raises the child

Thus, depending on how "parent" is defined, a child born of a surrogacy arrangement can be said to have as many as five "parents." The legal, ethical, and emotional consequences, if not anticipated and handled carefully, can be profound.

Questions of Custody Because donor insemination has been a widely accepted practice for many years, the law is clear on the point that a sperm donor cannot claim custody of a child born as a result of insemination with his semen. In the case of adoption, legal issues are also clearly defined— the birth mother gives up her parental rights to the child. In cases of surrogacy, however, a woman donates the egg and/or carries the pregnancy, and the law has been less clear.

In the late 1980s, two landmark cases received much media attention and are still affecting legal decisions in surrogacy cases today. As more of these cases are litigated in courts, precedents begin to be set upon which future decisions may be made.

Two Surrogacy Cases Because it involves not an anonymous donor but a woman who agrees to carry a pregnancy to term for the intended parents, surrogacy presents complex legal questions if disputes arise among the parties. The complexities are aptly illustrated by the cases of Baby M and JayCee B.

Baby M. William and Elizabeth ("Betsy") Stern, a New Jersey couple, wanted to enter into a surrogacy arrangement in order to have a child. Betsy Stern was a physician who had multiple sclerosis. She wanted to avoid carrying a pregnancy because, if her disease worsened—as it often does during pregnancy—she could find it difficult or impossible to continue in her profession afterward.

The Sterns found a willing surrogate in Mary Beth Whitehead, a married woman with two children of her own. The agreement signed by all parties was for Mrs. Whitehead to be inseminated with Mr. Stern's sperm. In return for giving up her parental rights to the child, Mrs. Whitehead would receive $10,000 plus reimbursement for her medical bills during the pregnancy.

After the baby was born, however, Mrs. Whitehead had a change of heart. She refused to give up the baby girl, whom the Sterns had named Melissa, to the Sterns. In response, the Sterns took legal action to gain custody of the baby, who was to become known in the news media as Baby M.

When a New Jersey court gave temporary custody of Melissa to Mr. Stern, Mrs. Whitehead fled with the baby, her husband, and her other two children. This act, together with threats she made to Mr. Stern in a taped phone conversation and the Whiteheads' lack of financial resources, contributed to the court's unanimous decision to award primary custody of Melissa to her biological father. Mrs. Whitehead retained parental rights and was allowed to visit Melissa under supervision for a specified frequency and length of time. Because Mrs. Whitehead retained parental rights, Mrs. Stern was not allowed to adopt Melissa.

JayCee B. In the case of "JayCee B." a Southern California couple, John and Luanne Buzzanca, arranged to have a child through a surrogate after years of unsuccessful treatment for infertility. The child would be born from an egg donated by one woman, fertilized with sperm from an anonymous donor, and implanted in another woman, who would carry the child to term for the Buzzancas.

In 1995, a month before the baby was born, Mr. Buzzanca filed for divorce and refused to pay child support. He took the position that because the child was not genetically his, he should bear no responsibility for supporting it. A lawyer for Mrs. Buzzanca argued that John had signed the contract agreeing to the surrogacy arrangement and that this made him the legal father. John claimed that he had signed the contract only to appease his wife, who was desperate to have a child before he left her.

The situation became more complicated when the surrogate, Pamela Snell, stepped forward and sued for custody of the baby girl, who by now had been named JayCee. Ms. Snell said that she had agreed to carry the pregnancy and deliver the baby with the belief that the child would grow up in a stable home with two married parents. She felt that Luanne Buzzanca had kept her impending divorce a secret until after the baby was born. Because Ms. Snell had carried the pregnancy to term, she felt that she was the baby's rightful mother.

More complications arose when the woman who had donated the egg came forward and claimed that the egg had been used without her permission. The sperm donor and his wife also spoke up, claiming that his sperm had been used without his permission. Both of these parties eventually dropped or reversed their claims, and neither one asked for custody of JayCee. Nevertheless, their appearance added confusion to the case.

Ms. Snell also eventually withdrew her claim, and in March 1998, a New Jersey court awarded custody to Luanne Buzzanca and required John Buzzanca to pay child support.

The court's reasoning in this decision was that "donating" a uterus to carry a pregnancy for another couple is no different from donating sperm to create a pregnancy. Both arrangements require the participation of someone who is not the intended parent of the child. The court also ruled that John Buzzanca could not consent to bringing a child into the world and then turn around and disclaim responsibility for that child. The judge concluded that "a deliberate procreator is as responsible as a casual inseminator." John Buzzanca has appealed the decision, and so the case is likely to linger in the court system for some time to come.

Getting Legal Advice Despite the infamous cases of Baby M and JayCee B., most surrogacy arrangements do not become embroiled in legal tangles. The key to preventing problems is to make every effort to ensure that all the parties involved understand and agree to the arrangement.

The cases of Baby M and JayCee B. have helped to make couples contemplating surrogacy more aware of the potential pitfalls of this option. Most surrogacy arrangements now include a contract to which all involved parties agree. Still, it is not clear whether these contracts are legally binding under any circumstances, or whether they can be nullified if the birth mother changes her mind about giving up her parental rights. Very few states have any laws to determine what is to be done in these situations.

Couples entering into surrogacy arrangements need clear understanding of the possible legal and emotional risks. It is usually advisable for both the surrogate and the intended parents to be represented by legal counsel to ensure that they fully understand the agreement into which they are entering. Many surrogacy programs also require psychological screening of potential surrogates. Attention to these issues will help everyone involved to understand the possible negative outcomes and to agree on how to handle them should they arise.

Social and Family Issues

Despite advances in medical knowledge, many still perceive infertility to be an embarrassing or shameful topic. Couples trying without success to conceive often feel isolated and alone, without a sympathetic ear among their friends and families. As one woman who has struggled with infertility says, "People have no problem talking about all the most intimate details of being pregnant and their child's birth. But just bring up the subject of *trying* to make those things happen, and suddenly there's a silence in the room."

It is difficult enough to deal with infertility without the additional burden of society's negative attitudes and clumsy reactions. To sort out these external sources of pain from your own feelings, it is helpful to consider the history of infertility over the last few centuries, which can reveal much about the source of attitudes that linger today.

Fertility and the Family from Past to Present

The discomfort and shame that often surround the subject of infertility can be traced back in part to religious and social attitudes and beliefs. For centuries, infertility was widely believed to be a "woman's problem." Not only were women expected to want children, but the cause of infertility was believed always to lie with them. A woman who felt no desire to become a mother was often thought of as unwomanly.

In addition, society's views of what makes a family or a household have shifted from communal arrangements to the more isolated structure of the nuclear family. This isolation has increased individuals' sense of independence, but it has also contributed to infertile couples' feelings that they are alone in their problems.

Changing Definitions For much of the past century, the "typical" American household has consisted of a married man and woman with one or more of their own biological children. The so-called "nuclear" family, however, is a relatively recent version of the American household.

In Colonial times, it was not uncommon for some children to be raised by adults other than their parents. Especially in large families with limited or modest means, younger children might be raised by childless relatives of the parents. In other instances, once they had reached a certain age, children might be sent to live in another household to study or to work as apprentices or servants. In their new households, they would typically receive food and lodging, and perhaps even a limited education. If they received an income for their labors, they often would send part or all of it back to their original families.

With the rise of democracy after the Revolutionary War, the social fabric of America began to change. Family structures shifted from communal households toward nuclear families. Women's roles became increasingly centered on the upkeep of the home and the raising of their own children, areas that were expected to be the primary focus of their lives.

Traces of the shift in the American household from communal arrangements to private, nuclear families are still very much in evidence. Today a high value is still placed on the nuclear family composed of a male and female couple with one or more of their own biological children. Other family arrangements, however, are becoming prominent, and mainstream society is beginning to recognize and accept them. This new shift is being brought about by social changes such as increased numbers of women working outside the home and larger numbers of divorced and single parents.

Today, a family can consist of many different groupings. "Blended" families result from divorce and remarriage. Families with adopted children are common, and women and men are choosing to become single parents through donor insemination and adoption. The high divorce rate in the United States has led to an unprecedented number of homes where one parent takes the primary or sole responsibility for children. And the "final frontier" may be couples who remain childless by choice but still think of themselves as a family.

Historical Religious Views In eighteenth-century America, most people of European descent believed that it was up to God whether or not a couple had children. If a couple remained childless despite their efforts to conceive, it was seen as "God's will." Why children were granted to some couples and withheld from others was a matter of divine mystery. So ingrained was this belief that efforts to treat infertility—most often by mid-wives using herbal remedies—were looked upon with suspicion as attempts to subvert Divine Providence.

Despite this religious view of fertility, the medical profession believed that infertility was almost always due to physical problems in the woman. A man was thought to be capable of fathering a child if he could achieve and sustain an erection. The cause of infertility, or "barrenness," was therefore almost always placed with the woman. This belief combined with prevailing religious views to create a double bind for women, who received the following message: "Infertility is God's will, and it is your fault." It was not a great leap to conclude that a woman's barrenness was a punishment from God. Even in today's relatively secular society, some women struggling with infertility feel that God is punishing them or controlling their lives.

The scientific and religious views of infertility as chiefly a "woman's problem" were even further compounded by social attitudes that a woman's proper role in life was to raise children, preferably her own. Thus society created a powerful source of shame and despair for a woman who could not become pregnant. She was held to blame, both physically and spiritually, as the source of the problem. The attitude dictated that she would never be a "true" woman or able to fulfill her role as a mother. Given these societal beliefs, it is easy to see why, especially for women, infertility was a subject clothed in shame.

For men, the subject of infertility is no less difficult. Like women, men generally set great store by their physical and sexual ability to produce off-spring. Once it was discovered that infertility could be caused by a male factor, many men suffered, and continue to suffer, blows to their self-esteem that are just as painful as those experienced by women. Since the subject of sex is so powerfully linked to procreation, many infertile men also feel that they are inadequate sexually as well.

Attitudes of Others Despite the many changes taking place in American attitudes about families, sex, and child-raising, many people still think of having children as the "default setting," the "norm." Most people still think of marriage and children as going hand in hand. Therefore, married couples without children may often be asked questions such as "When are you going to have a family?" by their family members and friends. It is interesting that this question asks not *if*, but *when*. An assumption is being made that it is something the couple *will* do—if not now, then in time.

Questions like these are so commonplace that they are probably not given much thought by most people. Although these remarks may be well-meant, they carry many expectations, such as:

- Everyone wants to have children.

- You are not a true family unless you have children.

- You cannot be truly happy, or fulfill your true destiny as a woman or man, unless you have children.

Such questions can be intensely personal for anyone, but for couples coping with infertility, they can be extremely hurtful. Understanding how society's views shape people's attitudes can help you to know how to respond to them without hostility. Some practical advice for dealing with others' well-meant but hurtful attitudes is offered below.

Dealing with Family, Friends, and Coworkers

The loss of something that is intangible, that cannot be touched or held, is still a loss. The loss of a child is heartbreaking, and the loss of a hoped-for child can be just as painful. But because infertility causes the loss not of a living child or a viable pregnancy, but rather of your hopes and wishes for one, it can be difficult for others to understand. Your loved ones are no doubt well-meaning, and may even be debating the most sensitive way to interact with you (for example, whether to invite you to a baby shower). Even so, they can still sometimes say or do things that add to your pain.

Even apart from other's remarks, many couples struggling with infertility find it difficult to attend family and social occasions where children are present. There are some things you can do, though, to help yourself through times like these. By controlling or minimizing the most painful aspects of these situations, you can begin to help yourself manage the emotions you are feeling.

Holidays and Social Gatherings Holidays, parties, and family gatherings can be stressful enough on their own, even without the issues raised by infertility. Often there is pressure at these times to enjoy yourself when you may be feeling far from joyful. Seeing happy families can underscore the absence of your own hoped-for children.

It will help you to get through these difficult occasions if you can try to identify in advance the things that are most likely to be upsetting. For example, it may not be a good idea to go to a party where there will be lots of children and pregnant women. This is not to say that you should always avoid such occasions. But it can be helpful to realize that you do have a choice. If you know that this tends to be a difficult issue for you, you can minimize your exposure to situations that are most likely to add to your pain, at least until you feel more ready to handle them.

Here are some suggestions to get you started on thinking about what your own difficulties are and how you might be able to minimize them:

- Before responding to an invitation, make it a habit to take some time to think about it first. Tell the person you will let them know in a day or two. Then mull it over for a bit and get in touch with how you feel about it, what it will be like, and what difficulties it may present. Don't automatically assume that you'll feel OK about it.

- Remember that you always have the option of saying no, and don't feel guilty about doing so. You are having a difficult time, and those who love you need to understand that. Don't cave in to guilt from family members or anyone else who may not understand what is best for you now.

- If you feel overloaded with too much time spent with other families and children, broaden your socializing a little to include unmarried friends and couples without children. Concentrate on developing more friendships with people with whom you feel you can "get away," instead of always having to be around children.

- On holidays, limit the time you spend with family members if you need to. It is normal to feel sad around other people's children when you have infertility problems—even if those children are your own nieces and nephews. Don't feel guilty about this. If it helps, plan to arrive at family gatherings when your exposure to children will be minimal, such as right before a meal or their bedtime. On holidays when gift-giving is a part of your family's tradition, consider arriving at a time other than when you know the children will be excitedly unwrapping their presents.

- If you shop during a holiday season, seek out smaller, quieter shops and boutiques instead of huge shopping malls crowded with families and children. Shopping by catalog or on the Internet can be a great alternative to braving the crowds.

- Finally, find your own ways of making holidays meaningful to you. You and your partner or friends can create new traditions instead of relying on old family ones. Use time away from your family to deepen friendships with the people who are best at offering you support.

None of this advice is meant to imply that you should shut yourself away completely from all situations involving children. Rather, it suggests that you may need to limit your exposure to these situations until you feel stronger and more able to handle them. Many couples facing infertility find that, with time, it becomes easier to be around children. But don't rush it.

Take the time you need to heal a little before throwing yourself into difficult situations.

Responding to Hurtful Remarks Communicating with family and friends can be one of the most difficult aspects of infertility. Many people have a hard time understanding the pain of infertile couples. They may say things without thinking or with the intention of helping, but that are nonetheless painful for you.

The first thing to remember is that *this person is not trying to hurt you.* If you value the relationship, you may need to help the person to understand what you are going through by being honest and direct without getting angry. If you care for the person, then you probably want her to know that although her remarks may be painful for you, this does not mean that you think she is a "bad" person. This can be a real challenge, but it helps to be able to anticipate some of the more common remarks so that you have some ideas about how to respond. Here are some examples that may help you to do that:

When are you going to start a family?

This question simply reveals that the person speaking is ignorant of your infertility problem. You may need to inform them of that fact without lashing out. It helps to assume that if they were aware of your problem they would never have asked you this question. Here's one way you could respond:

> I appreciate your interest, but the truth is that we're having a problem with infertility. It's really quite a painful experience, so I might not always want to talk about it. But thanks for your care and concern.

You should be glad you don't have kids—mine drive me nuts most of the time. I wish I had your freedom.

Believe it or not, the person who says this may actually be trying to comfort you. You need to let them know that their comment actually has the opposite effect. For example:

> I know that it's easy to feel that way when you are tied up in the day-to-day stress of raising a family, but I honestly look forward to having kids to complain about! What seems like freedom to you is really loneliness much of the time.

You could always adopt a child.

People often want to "fix" something when they see a friend in pain. This remark may be a clumsy attempt to offer a solution. You could say:

> I know that adoption is one possible option, but we have to first come to grips with the fact that we may not be able to have our own biological child.

Another suggested response is:

> I know you want to help, but I just need to feel bad about this for a while. The best way to help me right now is just to be there for me and be willing to listen.

So, any news?

What this person probably means is, "Is there any good news?" It's good to know that people care about your problem, but at the same time it can be hard to feel you have to keep everyone constantly updated. Sometimes you may not feel like going into the subject. You may need to let them know this:

> Thanks for asking. There's really nothing new I feel like talking about right now. I'll be sure to let you know if I need a sympathetic ear.

The situations you encounter on a daily basis may resemble or be very different from these examples, but the emotional response is usually the same. Try to let people know what you need from them in a way that is honest and direct without turning away those you care about. Understanding your own grief over infertility is necessary before you can communicate your needs to others in a way that they will understand.

Personal Issues

Without question, infertility is an emotionally charged issue for many couples. Finding that you may not be able to have children in the conventional way can be a source of pain for many reasons. Ultimately, though, it represents the personal loss of something dearly wished for.

It will help to sort out for yourself what this loss means to you—so that you can help others understand what you are going through, so that you can take steps toward understanding it yourself, and so that you and your partner can understand and help each other.

Grieving Your Loss

When we think of "loss," most of us think of losing something tangible, such as the death of a loved one. But the death of something that you cannot physically touch or see—such as a friendship, an imagined relationship with a child, or your hopes for a family—can be just as painful.

Any type of loss is really about the loss of an attachment or bond with something or someone. This is an important point to consider in understanding the pain of infertility: It can hurt just as much to have a bond broken with another person, such as by death, as it does to lose an attachment to a possibility—to hopes and dreams that have not yet come to be.

Grieving is a healing process, and the process has been recognized as having stages that most people experience in more or less the same way. The stages of grief are described in detail in Chapter 7 (page 180). Although knowing how most people typically grieve doesn't make the pain go away, understanding the grieving process can be a first step toward coming to terms with your very real loss.

With many types of loss, such as the death of a loved one, there is a single, isolated event over which the bereaved person grieves. In contrast, your pain over infertility is likely to be rekindled with new disappointments if tests and procedures do not have the desired results.

Thus, you might experience pain over, for example, a negative pregnancy test, a diagnostic finding, or an unsuccessful IVF cycle. Your pain and grief may be triggered anew by further disappointments, such as the second or third failed IVF cycle or the news that a different approach must be sought. For these reasons, it is quite common for those struggling with infertility to feel overwhelmed by the confusing mix of their emotions. This can make the grief process even more complicated and difficult to understand.

Personal Meanings of Infertility

Everyone experiences the pain of infertility in different ways, but some issues are common to many infertile couples. Some of these themes might be the loss of your hopes and dreams of:

- Being a parent

- Carrying on your family's history, traits, and traditions; seeing your family's physical and personality traits in a new human being, one that you helped to create

- Having something meaningful at the center of your life that is beyond yourself; devoting yourself to another being

- An investment in, and expression of hope for, the future

- How you imagined your life would be

In addition, and especially because of the societal pressures mentioned earlier, many women may feel guilty if they have put off childbearing for a career. Others feel that infertility is somehow "deserved" because they elected to abort a pregnancy in the past.

Infertility causes many couples to reexamine what they want or were expecting from being a parent. By looking at some of the specific components of your loss, you may find that the things that are most important are still within your grasp, even if you are not able to have your own biological child without ART.

Managing Your Emotions

Understanding your feelings is an important first step in working through the painful issues of infertility. You may not be able to make your feelings go away overnight, but you can acknowledge your emotions and find ways to manage them.

Reducing Stress Stress is always with us, but coping with infertility can be expected to intensify the stress that is a normal part of daily life. There may be no magic wand to wave in order to make the stress disappear, but there are things that can help you and your partner to be aware of your stress level and manage it as well as you can. If you don't do this, a sort of vicious circle can develop, in which stress erodes your ability to function, which in turn tends to add to your stress.

The key to dealing with stress is not so much to get rid of it—that may not be possible—but rather to find ways to manage it. You might be surprised at some of the simple things that can help reduce your stress level. They mainly involve looking after all aspects of yourself—not just your physical health, but your mental and spiritual health as well.

Taking Care of Your Body During times when you feel depressed, angry, or upset, it is easy to lose your motivation to take care of yourself physically. But these are the times when it is more important than ever to be sure that you are eating a healthy diet and getting enough rest and exercise. If you are having trouble sleeping, then just rest when you can during the day so that you are not completely worn out.

Starting or keeping up with a regular exercise program can go a long way toward reducing stress, depression, and anxiety. It can even help with insomnia, thereby creating a "beneficial circle" of its own: Exercising can help you to sleep better, and sleeping better boosts your energy level for exercise as well as other activities.

Taking Care of Your Mind and Spirit Taking care of yourself physically is important, but of course your body is more than a machine. Your mental health is also important. Keeping up your spirits is key to helping yourself during difficult times.

Being with other people who care about you and understand your feelings is an important element of staying mentally healthy. But you also need time to yourself, during which you can do the things you really enjoy, that are just for you. Solitude does not always equal loneliness; it can also be a time to reflect and get in touch with how you are feeling. You may want to make a stronger connection with your church, if you belong to one, or with other aspects of your spiritual life.

Take a walk in the woods, get up to see the sunrise in a special place, sit quietly with your favorite music on the stereo, or spend time doing, or rediscovering, something you really love—painting, singing, dancing, cooking, even playing a sport or riding a horse. Find the things that make you feel most connected with yourself and the world, and make time to spend on them on a regular basis.

Communicating with Your Partner Without question, there are differences in the ways people communicate and express themselves. If you and your partner are coping with infertility, communication may seem impossible at times. Ironically, it is during times of stress that communication becomes more important—as well as more difficult—than ever.

In addition to their differing communication styles, people in general also have different ways of dealing with stress and grief. This can be especially true of women and men. These differences can contribute to the problems between a couple and make them worse. It's easy for two people to become shut off from one another because they do not understand each other and do not want to contribute to each other's pain.

Try not to let this happen. Keep the lines of communication open, even—especially—when it seems the most difficult.

Look at your coping styles as two different languages. For each of you, your job is to translate or "decode" what the other is trying to express. It will take practice, and you will make mistakes. Do not hold these misunderstandings against one another. Forgive each other and move on—do not give up hope, and do keep trying.

A typical example of two different coping styles is when one partner is open and expressive and the other is restrained and controlled. Stereotypes usually ascribe the former style to women and the latter to men, but these roles can just as easily be reversed.

You and your partner can benefit from being aware of each other's differences and decoding each other's signals. If you just cannot seem to overcome the problems, seek out a qualified therapist, preferably one with experience in dealing with infertility issues. You may be able to get a recommendation from a friend. Or consult the resources listed at the end of this chapter for support groups that can point you toward the resources you need.

HANDLING COMMUNICATION DIFFERENCES

Chris has always tried to keep a stiff upper lip when faced with adversity. Now that he and his partner, Tracy, know that they have an infertility problem, he feels that remaining strong is the best way to help and protect her. His stoicism, however, is usually mistaken for indifference or coldness by Tracy, who is more comfortable expressing her emotions openly than keeping them inside.

It may seem strange to Tracy, but Chris's outward control is actually a way of showing how much he cares about her. When he feels helpless and doesn't know what to do to make her pain go away, Chris assumes an air of steely resolve in order to convey strength to Tracy.

Of course, this usually doesn't work. When Tracy is tearful, angry, depressed, or scared, she wants warm, loving, and sympathetic words from Chris, and to feel his arms around her. She feels pushed away by Chris's apparent indifference.

What Tracy needs to remember is that her partner loves her a great deal and is dealing with the pain in the best way he knows how. She must let him know what she needs, but she must also allow Chris to deal with his feelings in his own way. It will help if Tracy can decode Chris's restraint as an expression of his pain, and offer whatever comfort she can.

On the other hand, matters are made worse for Chris when Tracy so easily expresses her feelings. Chris often feels that, because his partner is so distraught, he could not possibly expect her to give him the support he needs. He feels that to express his own feelings to Tracy would be just a further burden to her.

However, it is important for Chris to do some decoding here, too. It may seem contrary to logic, but the chances are great that if he allowed himself to break down in front of Tracy, this would actually be reassuring to her. She is likely to feel that Chris truly does understand her and that his feelings are a validation of her own.

In time, Chris and Tracy are able to talk this through. They both come to feel that they understand one another better. Eventually, they are able to map out some middle ground—a place where each can deal with their feelings in their own way while still giving the other as much love and support as they are able.

Tracy makes up her mind that she will try to remember that Chris's emotional distance almost surely masks feelings of pain, grief, and sadness. This is his way of dealing with the pain on his own terms. She may need to allow Chris to be withdrawn sometimes. But together they agree that they will not let these silences persist too long. Eventually—in an hour or a day—they promise that they will talk to each other about what each of them is feeling.

For his part, Chris realizes that Tracy's emotional nature does not mean that there is no room for his own feelings. He sees that sometimes he has used Tracy's expressiveness as an excuse not to talk about his own feelings, which he has always found difficult to do. He decides that even when she seems upset, he will try to remain open about how he feels, and even to cry in front of her if he wants to.

Sexual Intimacy Infertility can have a huge impact on a person's sexual self-image. Both women and men can experience feelings that their failure to procreate implies that they are somehow deficient sexually—not a "true man" or a "true woman." For an infertile couple, sex can become associated with pain and loss. It can become impossible for either partner to enjoy sex in a caring and intimate way when it serves only to remind them of the fact that they have not been able to conceive.

In addition, for many couples undergoing infertility treatment, sex can begin to seem like a chore. Focusing so closely on the woman's menstrual cycle and signs of ovulation, taking daily basal body temperature readings, and timing intercourse can all have a way of taking the spontaneity out of a couple's sex life. It is quite natural for either or both of them to begin to dread sexual encounters because their hopes for a pregnancy have been dashed so often. It is one of the ironies of infertility that this area of a couple's life together, which should be the source of closeness, intimacy, and shared joy, often becomes so closely linked to the source of their pain that it serves only to drive them apart.

Like any relationship problem, sexual difficulties can be eased through open and honest communication. If your time outside of the bedroom is spent growing closer and feeling more connected emotionally, your sexual problems will be eased. In addition, there are many professional therapists who specialize in helping couples with this problem. A good fertility clinic should be able to refer you to one.

Facing Infertility Alone Of the estimated four million births that occur each year in the United States, more than 400,000—greater than 10 percent—are to unmarried women. It is not known exactly how many of these births are planned and intentional, but it is true that increasing numbers of women are choosing single motherhood.

A woman who has consciously chosen to become a single mother faces a unique set of difficulties. Her biggest problem, contrary to what many people might expect, is not necessarily her lack of a partner. Many of these women have very supportive networks of friends and family and actually feel less isolated than women who have chosen the "traditional" route of staying home alone with their babies. Rather, many single mothers report that it is other people's attitudes about their choice that poses the greatest challenge.

If you are a woman who has decided to become a single parent, you are grappling with additional challenges if you have an infertility problem. It is more important than ever to take care of yourself physically, emotionally, and spiritually during this time. Some women in this situation appoint one or two very close friends, or perhaps a close relative such as a sister, to act as their primary source of empathy and support. This person may also be the one chosen to attend your delivery should you succeed in carrying a pregnancy to term.

Take some time to sit down and make a list of the people whom you feel you can call upon during particularly stressful times, and ask them directly if they would feel comfortable in this role. In addition, you can seek out support groups for single parents in your area through the recommendations of your doctor or counselor.

Finding Help

Many potential problems and challenges can attend infertility, even apart from any medical or health problems that are uncovered. You cannot possibly anticipate all of these trouble areas, but you can identify some of the ones that are most likely to occur in your particular situation. Thinking about what steps to take in advance, before problems arise, will be enormously helpful in dealing with them.

It is also important to know your own limits—physically, financially, and emotionally—in terms of what you can handle during the course of your infertility treatment. With so many dramatic advances being made in today's changing world of reproductive medicine, it is only natural to want to try every possible option. Remember, though, that you do not have to exhaust all the technological possibilities simply because they can be done. People vary in their ability to face the complicated problems linked to these technologies. For example, for one person or couple, having their own biological child may be a high priority. For others, this may be less important than experiencing the pregnancy firsthand. For yet others, the raising of a child, regardless of whether it is genetically related to them or was carried and delivered by the woman herself, is the most important factor.

From time to time, look back at the suggestions in Chapter 3 (page 38) to clarify your own priorities. Think about the potential legal, practical, and personal problems of each choice, and which of these seem like they would be the biggest obstacles for you.

Finally, you need to know when to stop if treatment has been unsuccessful. Many infertile couples decide that they do not want to handle any more disappointments and no longer want their entire lives to be centered around this one problem. If you reach the point where you have exhausted the financial, physical, and/or emotional resources you need to cope with infertility treatment, it is time to think about stopping. Chapter 8 addresses this topic and offers suggestions for finding the next steps that feel right at this point in your life.

Suggested Resources for Dealing with Infertility

The **American Society for Reproductive Medicine (ASRM)**, the author of this book, is a voluntary nonprofit organization devoted to advancing knowledge and expertise in reproductive medicine and biology. ASRM fosters patient understanding and involvement in reproductive medicine through the Patient Information Series of booklets and a variety of other information sources that are free to the public.

> ASRM
> 1209 Montgomery Highway
> Birmingham, AL 35216-2809
> (205) 978-5000
> E-mail: asrm@asrm.org
> Website: http://www.asrm.org

RESOLVE is a national nonprofit organization that, for more than twenty years, has assisted people in resolving their infertility by providing information, support, and advocacy.

RESOLVE
1310 Broadway
Somerville, MA 02144-1779
Business office phone: 617-623-1156
National HelpLine: 617-623-0744
E-mail: resolveinc@aol.com
Website: http://www.resolve.org

Single Mothers by Choice (SMC) was founded in 1981 by Jane Mattes, C.S.W. and psychotherapist. SMC is devoted to providing information and support to single mothers, as well as those contemplating or trying to achieve single motherhood.

SMC
P.O. Box 1642
Gracie Square Station
New York, NY 10028
(212) 988-0993
E-mail: mattes@pipeline.com

The **International Fertility Center** provides coordination of medical, psychological, educational, and legal services for clients facing infertility problems.

International Fertility Center
2601 East Fortune Circle Drive
Suite 102B
Indianapolis, IN 46241
(317) 243-8793
E-mail: interfert@iquest.net
Website: http://www.fertilityhelp.com

Online Resources

Child of My Dreams is an online resource for individuals and families facing infertility and considering adoption.

E-mail: childdrm@child-dream.com
Website: http://www.child-dream.com

FertiNet disseminates information among professionals, researchers, and patients in the field of assisted reproduction.

E-mail: info@ferti.net
Website: http://www.ferti.net

Parents Through Assisted Reproduction (PTAR) maintains a listserv and online support center, including an e-mail newsletter.

E-mail: PTA-L-team@surrogacy.org
Website: http://www.surrogacy.com/online_support/pta/index.html

The American Surrogacy Center (TASC) provides support and information on a variety of subjects related to surrogacy, including legal, medical, psychological, and personal issues. TASC's website also provides links to discussion groups and articles on legal issues related to ART.

E-mail: TASC@surrogacy.com
Website: http://www.surrogacy.com

Books

What to Expect When You're Experiencing Infertility: How to Cope with the Emotional Crisis and Survive, by Debby Peoples, Harriet Ferguson, C.S.W., and Alice P. Domar. W. W. Norton and Co., 1998.

Healing the Infertile Family: Strengthening Your Relationship in the Search for Parenthood, by Gay Becker, Gaylene Becker, and Robert D. Nachtigall. University of California Press, 1997.

Sweet Grapes: How to Stop Being Infertile and Start Living Again, by Jean Carter and Michael Carter. Perspectives Press, 1998.

Taking Charge of Your Infertility, by Toni Weschler, M.P.H. HarperPerennial, 1995.

7

Pregnancy After Infertility

If you have finally become pregnant after infertility treatment, you probably have many questions about what to expect. If this is your first pregnancy, the usual feelings common to most new parents-to-be—excitement, anticipation, and wonder of a new life—may also be mixed with feelings of fear, anxiety, or even dread. Such a blessing, so hard-won, may seem too good to be true, and you may fear losing it.

Knowledge is the best defense against anxiety and fear. This chapter will help you to understand the risks in a pregnancy conceived with assisted reproductive techniques (ART). It will tell you what you can expect in terms of prenatal care, testing, and delivery options. This information will help you to understand how your pregnancy may—or may not—differ from one that occurs without technical assistance so that you will know how best to take care of yourself and of the new life growing inside you.

Mental and Physical Adjustments

Pregnancy is always a time of adjustment and transition. When it occurs after a period of infertility treatment, making these adjustments can be especially challenging.

Dealing with Anxiety

From worrying about monthly cycles and hoping for good news, you must now shift to focusing on the reality of bringing a new life into the world. It is understandable for both partners to have mixed emotions. The mother may feel that her condition is fragile, and she may be truly fearful that she will do something to jeopardize her child. The father is probably also worried, but may not want to annoy his partner by asking how she feels at every moment. Communication can break down as both partners strive to express their pleasure and hide their fears.

Knowing that your feelings are normal and understandable is the first step to taming fear and anxiety. Communication between partners is also key. Chapter 6 suggested ways to improve your communication skills. These techniques can continue to help you through the nine months of pregnancy.

Remember that your health care provider can reassure you by giving you a realistic picture of the risk of miscarriage during your pregnancy and by suggesting ways to lower the risk. When you learn all you can and communicate *all* your feelings to your partner and your doctor, eventually you will find yourself more able to relax.

 SHARING THE NEWS

Many couples wonder when to share the news of their pregnancy with families, friends, and coworkers. Some couples, particularly those who have experienced a miscarriage in the past, prefer to wait until the pregnancy advances before informing others. Others want to spread the news right away. It is best not to assume that your partner feels the way you do about this decision.

Talk about it and decide together before taking action. Are you likely to want the support of others from the very start? Or would you rather know that the pregnancy has a greater chance of continuing before sharing the news? Mulling over questions like these will help you and your partner make the decision that is right for both of you.

Managing Symptoms

Many infertile women have watched as the pregnancy of a friend or family member has progressed. When you find yourself carrying a child, you may be alert to every twinge and ache, and, for reassurance, you may constantly want to compare your symptoms with those of a "normal" pregnancy. It helps to know what daily symptoms to expect.

Common Side Effects of Pregnancy The common signs of pregnancy are just as likely to occur in an ART pregnancy as in unassisted pregnancies. Although these side effects are not signs of danger to the fetus, they can make you uncomfortable and anxious. Most will taper off by the second trimester. During the first trimester, however, you are likely to experience:

- Nausea and vomiting

- Fatigue

- Frequent urination

- Breast tenderness

- Headaches and dizziness

Your doctor can suggest ways to ease these symptoms and make yourself more comfortable. He or she can also tell you how to know when they are normal side effects of early pregnancy and when they may be cause for concern.

Ovarian Hyperstimulation Syndrome In addition to the usual symptoms of pregnancy, women with ART pregnancies may experience ovarian hyperstimulation syndrome (OHSS). This condition causes the ovaries to become swollen and painful. It is fairly common among women who have had controlled ovarian hyperstimulation (COH) for the purpose of producing multiple follicles for ART.

The symptoms of OHSS are usually mild to moderate, but in some cases they can be severe, as in women who have conceived with the aid of injectable ovulation drugs. In these women, OHSS may occur immediately after ovulation and may recur more severely in very early pregnancy.

In most cases of OHSS, the symptoms include:

- Bloating

- Nausea

- Vomiting

- Shortness of breath

- Reduced urinary output

- Weight gain of two pounds or more in a twenty-four–hour period

- Abdominal pain

- Loss of appetite

The symptoms can be treated with rest, analgesic medications, and fluids. In more severe cases, fluid may build up in the abdomen and chest.

When this happens, there is a risk of blood clots forming and other medical problems that require hospitalization.

OHSS generally subsides after several weeks of pregnancy. You should inform your doctor if any of the symptoms described here become severe.

Care of the ART Pregnancy

A pregnancy achieved with ART usually requires more than the usual amount of monitoring during the early part of gestation. Like any pregnancy, a program of prenatal care should be planned for the ART pregnancy. This plan will include making the transition from your infertility specialist to an obstetrician or other maternal–fetal medicine specialist.

Diagnosis and Monitoring

The type and frequency of monitoring that is most appropriate for an ART pregnancy will depend on several factors, including the mother's medical history, the type of fertility treatment she received, and whether she has had a previous miscarriage or an ectopic pregnancy.

Confirming Pregnancy Because a woman's menstrual cycle is closely monitored during infertility treatment, when pregnancy occurs, it is usually noted immediately. The physical symptoms alone usually alert a woman to the possibility of pregnancy, and lab tests confirm the diagnosis.

Physical Signs and Symptoms The most common sign of early pregnancy is a missed menstrual period. Sometimes light bleeding or spotting—much lighter than a normal menstrual flow—occurs around the time that a woman expects her period to begin. This is known as *implantation bleeding*, and it may occur when the fertilized egg implants in the lining of the uterus.

Other symptoms of early pregnancy include:

- Breast enlargement and nipple tenderness

- Nausea and vomiting

- Frequent urination

- Fatigue

- Increased vaginal discharge

- Headaches

- Mild cramping

All of these changes are normal signs of early pregnancy. Bear in mind, however, that progesterone supplementation, often used in ART, mimics many of these symptoms, and may even delay menses. Other symptoms may be warning signs that should be reported to your doctor right away. Call your doctor if you experience:

- Abdominal pain

- Severe cramping

- Any vaginal bleeding in excess of spotting

- Fever or chills

- Persistent vomiting

- Painful urination

Vaginal bleeding is common in early pregnancy. Some research suggests that this symptom may be more common in ART pregnancies. It is not always a sign of a serious problem, but it could indicate an ectopic pregnancy or an impending miscarriage.

Lab Tests Apart from physical symptoms, the results of lab tests are also used to make the diagnosis of pregnancy. Blood samples will be drawn to detect the levels of the hormones human chorionic gonadotropin (hCG) and progesterone. (See Chapter 5 for a refresher on the role of these hormones during pregnancy.) The results of these tests, together with ultrasound monitoring and physical changes, help to confirm pregnancy, and also give some prognostic information.

Ultrasound Monitoring Five to six weeks after a woman's last menses (or one to two weeks after a missed menstrual period), ultrasound is usually performed to detect the beating heart of the fetus. A vaginal ultrasound transducer, a slender device inserted into the vagina, will probably be used. The sound waves emitted by the transducer are bounced off the structures inside the woman's abdomen and transformed into images that can be viewed on a monitor. With this procedure, the doctor can confirm that the pregnancy is inside the uterus, and can detect how many embryos are present. As in spontaneous (non-ART) pregnancies, ultrasound may also be used later in an ART pregnancy to monitor the developing fetus. After about ten weeks, an abdominal ultrasound scanner is used.

A recently developed ultrasound technique provides three-dimensional images of the fetus. This is a new technology that is not yet widely available. Your doctor can give you more information about 3D ultrasound and its use in pregnancy.

Special Aspects of Prenatal Care

Some special considerations are given to ART pregnancies during the nine months of prenatal care, but for the most part you will need the same level of care as someone with a non-ART pregnancy. ART pregnancies usually differ because of the effects infertility treatment may have early on.

Transition to Obstetrical Care At some point after pregnancy is confirmed, a woman's care usually is transferred from her fertility specialist to an obstetrician. Many obstetricians have experience in managing the pregnancies of women who conceived through ART and are knowledgeable about the special aspects of prenatal care for these patients.

If there are special risks in a woman's pregnancy, the obstetrician will probably recommend a *perinatologist,* or maternal–fetal medicine specialist. This is a physician who has fulfilled all the requirements for an obstetrician-gynecologist and also has additional training and experience in managing pregnancies that are at special risk for complications.

Hormonal Support Some women with ART pregnancies are found to have a lower-than-normal production of the hormone progesterone. These women may be prescribed hormones in early pregnancy to support the development of the embryo. Natural progesterone is identical to the hormone produced by the placenta and ovaries during pregnancy. Unlike synthetic progestins, natural progesterone has not been linked to an increased risk of birth defects.

Progesterone is also often used after ART procedures such as IVF, GIFT, and ZIFT. Women who have had recurrent miscarriages also use progesterone. Progesterone supplements are usually continued through part or all of the first two months of pregnancy.

Complications in ART Pregnancies

Whether ART pregnancies are at a higher overall risk for complications is still unknown. However, infertile couples usually start out with a higher risk for many of the complications of pregnancy in the first place. The problems that bring these couples to seek infertility treatment, such as the age of the woman or tubal scarring or blockage, are many of the same problems that also place them at higher risk for pregnancy complications. For this reason, directly comparing their pregnancies with those of fertile couples does not necessarily yield a realistic picture.

Clearly, though, the presence of multiple fetuses is more common in ART pregnancies. And multiple pregnancy itself increases the risk for both the mother and the fetuses before, during, and after delivery. In research done so far, it is not clear that an ART pregnancy with a single fetus (commonly called a *singleton pregnancy*) is at any higher risk than a non-ART singleton pregnancy. Fortunately, most ART pregnancies, whether multiple or not, have the advantage of being very closely monitored, so that any problems that do arise are more likely to be found early.

Multiple Pregnancy

In most ART procedures, pregnancies are conceived with the aid of "fertility drugs"—usually either clomiphene or gonadotropins. Stimulating the ovaries to produce multiple follicles increases the chances that at least one egg will be fertilized. In ART procedures such as IVF, GIFT, and PROST—all of which typically include COH—several embryos are usually placed in the woman's uterus in order to increase the chances that at least one will survive and be carried to term.

Obviously, all of these procedures raise the odds for the fertilization and survival of multiple eggs and embryos. Twins occur in 1 to 2 percent of all spontaneous pregnancies, and triplets are even more rare. These rates are much higher in ART pregnancies. How much higher they are depends on, among other things, what type of medications are used for COH. With the use of clomiphene citrate, the risk of multiple pregnancy increases to 5 to 8 percent, the vast majority of which are twins. With gonadotropins, the risk is much higher, at 20 to 25 percent, about one-third of which are triplets or more.

Risks Associated with Multiple Pregnancy The risks linked to multiple pregnancy increase with the number of fetuses.

Preterm labor and stillbirth are markedly increased in a pregnancy with three or more fetuses, as is the rate of miscarriage. Some studies have found the miscarriage rate to be more than 10 percent in triplet ART pregnancies, more than 20 percent for quadruplets or quintuplets, and up to 50 percent for sextuplets. For this reason, many fertility programs make it a policy not to transfer more than three to six fertilized eggs during an IVF procedure. When pregnancy is conceived spontaneously via COH, however (such as with intrauterine insemination), there is less control over how many embryos implant in the uterus.

In addition to the risks posed to the fetuses in a multiple pregnancy, the mother is also more likely to have certain complications, such as circulatory problems. She is also likely to be more uncomfortable than if she were carrying a single fetus, because her uterus is much larger and heavier with multiple fetuses. Women with multiple pregnancies are often required to remain on bedrest during the pregnancy—for weeks or even months.

Despite the risks, many women do carry multiple ART pregnancies to term and deliver healthy babies. Early detection of multiple pregnancy through ultrasound monitoring and other tests can help to solve problems before they become serious.

Hypertension Multiple pregnancy increases the risk of *pregnancy-induced hypertension* (PIH), or high blood pressure, in the mother. (Another term for this condition is *preeclampsia*). High blood pressure in general becomes more common with age. Because many women undergoing ART are in their 30s and 40s, PIH may be more common among these women. As with other pregnancy complications, though, it is not clear that ART itself directly increases the risk of PIH, independently of the risks conferred by age and multiple pregnancy.

If unchecked, PIH can lead to serious problems in the mother, including kidney or liver damage, bleeding problems, or seizures. (PIH that has led to seizures is called *eclampsia*.) PIH can mean that the fetus does not get enough oxygen or nutrients from the placenta (the thick pad of tissue inside the fetal sac that provides nutrients and carries away waste). This, in turn, places the fetus at risk for growth retardation and prematurity.

The management of PIH depends on its severity. Less severe cases may require only close monitoring and frequent office visits to the obstetrician. In more serious cases, the woman may be put on bedrest for weeks or even months. Severe PIH may require hospitalization and medication to prevent seizures and control blood pressure. Finally, early delivery may become necessary if the fetus or mother appears to be in danger.

Regular monitoring of blood pressure during pregnancy can identify PIH early, when it is more likely to be mild and is more easily kept under control. Regular blood pressure monitoring is particularly important when it is known that the woman is carrying more than one fetus.

Preterm Labor and Delivery Labor that begins before the end of the thirty-seventh week of pregnancy is considered preterm. Before this time, the fetus is less likely to be fully mature and able to survive on its own outside the uterus. Preterm labor and delivery are more common in multiple pregnancies, although the exact reason for this increased risk is not known with certainty.

Preterm babies are at risk for a host of complications both during and after birth. They are more likely to be in a breech position at the time of delivery. Delivery of a breech baby is more difficult and is more likely to require a cesarean delivery. Preterm babies are also more likely to have low birth weight—weighing less than five-and-a-half pounds at birth. Because of their decreased body fat, these babies have a harder time keeping a normal body temperature. And because their lungs are less likely to be fully mature at birth, they are at risk for a condition called *respiratory distress syndrome* (RDS). Their other internal organs may also be underdeveloped, and they may have difficulty swallowing.

Depending on their condition at birth and how underweight they are, many preterm babies need to stay in a *neonatal intensive care unit* (NICU) for a period of time after birth. They may need a respirator to help them breathe, and may need to be fed through a tube. Typically, these babies stay in the NICU at least until the date on which their birth was expected to occur. Those with more severe problems may need a longer stay.

Preterm labor may be signaled by a number of signs and symptoms:

- A vaginal discharge that is greater or different than usual

- A feeling of pressure in the lower abdomen or pelvis

- A low, dull backache

- Uterine cramping, contractions, or tightening

If a pregnant woman shows signs of starting labor early, she may be advised to limit or refrain from many physical activities. In some cases she may need to stay on bedrest, possibly in a hospital, for the remainder of her pregnancy. There her condition will be closely monitored, and if labor begins, steps may be taken to halt it. If one or more of the fetuses appears to be in danger, doctors may induce an early delivery.

Fetal Growth Retardation Infants who are born small for their gestational age—the length of time they have spent in the uterus since conception—have a higher risk of complications and death in the period after birth. These infants are said to have *fetal growth retardation* or *intrauterine growth retardation* (IUGR). This condition is distinct from preterm birth in that a growth-retarded baby may be born at the expected time but still be smaller than normal. Preterm babies may also show signs of IUGR if they are even smaller than what is normal for their gestational age.

IUGR of one or more fetuses is more common in a multiple pregnancy. In twin pregnancies, it is not unusual for one twin to be growth-retarded and the other to be of normal size. After birth, growth-retarded babies may be more likely to remain physically and possibly mentally less developed than normal infants.

Growth retardation is sometimes detected during pregnancy by the use of ultrasound. If the diagnosis of IUGR is confirmed, the woman is likely to need more frequent monitoring of her pregnancy and of the fetuses. In cases where one or more fetuses appears to be at risk, an early delivery is usually recommended.

Breech Presentation In most pregnancies, the fetus moves into the normal birth position several weeks before delivery. This normal position is called a *vertex presentation*—in which the baby's head is positioned downward, ready to emerge first. In a *breech presentation*, the baby's position is reversed, with the feet or buttocks pointed downward. In a multiple pregnancy, breech presentation of at least one of the fetuses is common. Delivery of breech babies, while safe, is technically challenging. When the second twin is breech, a safe vaginal delivery is likely.

Cesarean Delivery When delivery through the vagina is judged to be too risky—as is often the case in a multiple pregnancy—a cesarean delivery is performed. In a cesarean delivery, the baby is born through a surgical incision made in the mother's abdomen and uterus. Like any major operation, cesarean delivery places the mother at risk for complications from anesthesia and excessive blood loss. However, the vast majority of cesarean deliveries are successful.

Cesarean delivery is usually performed with the mother under a regional anesthetic, which blocks pain from the waist down. The surgeon makes an incision into the woman's abdomen and uterus and delivers the babies through this opening. Cesarean delivery requires a longer hospital stay for the mother and a longer recovery period than vaginal birth. Complications for the mother from cesarean birth can include infection, excessive blood loss, blood clots, and injury to nearby organs and structures. In the absence of problems like these, it can still take up to three weeks before the mother can resume a normal activity level.

Postpartum Hemorrhage Hemorrhage, or excessive blood loss, immediately after delivery can be life-threatening for the mother. *Postpartum hemorrhage* (PPH) is classified as either early or late. Early PPH occurs within the first twenty-four hours after delivery, whereas late PPH can occur from twenty-four hours to six weeks afterward.

PPH is more common with a multiple pregnancy, partly because of the increased likelihood of cesarean delivery. Another reason is that the uterus is stretched much more than in a singleton pregnancy. This stretching tends to weaken the muscle of the uterine wall, so that it is slower to return to its normal tone after delivery.

PPH may require a hospital stay so that the mother can receive a blood transfusion and monitoring. In some cases, surgery may be needed to correct the cause of the bleeding.

SELECTIVE REDUCTION IN MULTIPLE PREGNANCY

A multiple pregnancy with three or more fetuses is at far greater risk than a singleton or even a twin pregnancy. For this reason, if a woman is found to be carrying triplets or more, she and her partner might be offered the option of *selective reduction.* This procedure is sometimes also called *fetal reduction, selective birth,* or *multifetal pregnancy reduction.* It may be done to improve the chances of survival of some of the fetuses, or to terminate a fetus or fetuses with serious birth defects that would not allow them to survive.

Selective reduction is generally done in the first trimester (up to twelve weeks) of pregnancy. In the most commonly used method, ultrasound scanning is used to guide a long, thin needle through the woman's abdomen and into the uterus. A solution containing potassium chloride is injected into the fetuses chosen—usually those most easily reached. This solution causes the hearts of those fetuses to stop beating.

When selective reduction is performed, the remaining fetuses are at a slight increase in risk for miscarriage. For this reason, before choosing to perform this procedure, a couple and their doctor must be reasonably certain that the risk of miscarriage of the remaining fetuses is not as great as the risk of death of all of the fetuses if the procedure is not performed.

For obvious reasons, selective reduction is a highly charged emotional and ethical issue for many people. Couples faced with this option must weigh their decision very carefully. First and foremost they must ask whether their personal, moral, and religious beliefs allow them to consider this possibility. Then, if the procedure is not done, will they be willing to risk losing all of the fetuses? Can they live with the choice of terminating some of the fetuses in order to allow the others to live? What are their moral responsibilities to the potential lives they have created with the help of technology? These are just some of the issues at stake in weighing this difficult decision.

Most clinics try to prevent a couple from ever having to make this choice by limiting the number of embryos transferred during IVF. If you are faced with this extremely difficult decision, talk it over with your doctor. It may also be helpful to discuss it with a counselor or clergy person.

Pregnancy Loss

The rate of pregnancy loss may be higher in ART pregnancies, partly because more of these pregnancies are multiple. Other reasons for increased risk include the same factors that contribute to infertility, such as advanced maternal age.

In addition to *miscarriage*—the loss of a pregnancy before twenty weeks—the rate of ectopic pregnancy is also higher in ART pregnancies. In an ectopic pregnancy, the fertilized egg implants outside of the uterus, most often in a fallopian tube.

Miscarriage It is estimated that as many as 25 percent of all pregnancies end in miscarriage. Although the rate of miscarriage in ART pregnancies may be higher than in other pregnancies, this cannot be stated with certainty. Since spontaneous pregnancies often terminate before they are ever recognized, and since ART pregnancies are usually diagnosed quickly, early recognition may be one reason for the perception that ART pregnancies may have a higher miscarriage rate.

Most miscarriages are caused by a random problem with the growing fetus and are not related to the mother's health. Recurrent miscarriages, however, may be due to factors of the mother's health, such as:

- Abnormalities in the shape of the uterus, such as might occur in women exposed to DES

- Uterine fibroids

- Hormonal disorders

- Age over 35 (the risk of miscarriage increases with age, probably because of the aging of a woman's oocytes)

- Problems with the immune system

- Cigarette smoking

Biochemical Pregnancy Some pregnancies are lost because, despite all the signs of pregnancy, no embryo has actually formed from a fertilized egg. In these cases, a woman has all the normal signs and symptoms of pregnancy, including a missed period. Her blood tests, too, show hormone levels that normally would confirm the diagnosis of pregnancy. When an ultrasound exam is done, however, no embryo is found inside the gestational sac that has implanted in the uterine lining.

This occurrence is sometimes called *biochemical pregnancy* because hormone levels seem to indicate a pregnancy although none is actually present. Sometimes it is called *blighted ovum*, because it is thought to be due to a defect in the egg that was fertilized. Biochemical pregnancy seems to be more common in ART pregnancies, possibly because many women who conceive with ART are in their 30s and 40s, when defective eggs and miscarriage are more common.

Ectopic Pregnancy The risk of ectopic pregnancy may be higher in women who have conceived through ART. By far, most ectopic pregnancies—about 95 percent—occur in a fallopian tube, although they may also occur in the cervix, ovary, or the abdominal cavity. A tubal ectopic pregnancy may occur anywhere along the length of the fallopian tube. When it is in the end of the tube near an ovary, it is referred to as *ampullary*; one in the middle portion of the tube is called *isthmic*; and one in the part of the tube that opens into the uterus is called *interstitial* (see Figure 7.1). Because the fallopian tube is not designed to accommodate a growing fetus, the risk of an ectopic pregnancy is that the tube may burst, leading to life-threatening complications.

Women whose tubes are scarred or damaged are at higher risk for ectopic pregnancy. In fact, as many as half of ectopic pregnancies are thought to be related to tubal disease. Tubal damage may result from prior surgery on the tubes, such as an operation to reverse tubal ligation; from a past infection, such as pelvic inflammatory disease; from endometriosis; or

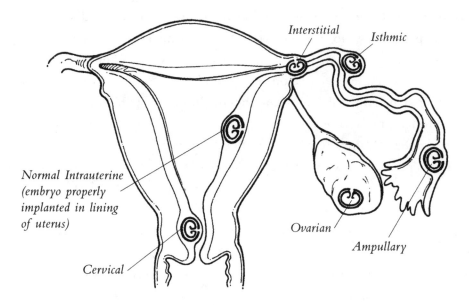

Figure 7.1: Ectopic pregnancy.

from a past ectopic pregnancy. (See Chapter 4 for more details on infertility due to tubal damage.)

The earliest signs of an ectopic pregnancy are the same as those for a normal pregnancy—missed periods, breast tenderness, and possibly spotting around the time of the expected period. If blood tests confirm that a pregnancy has occurred, the first indication that it is ectopic is often pain in the lower abdomen and/or irregular bleeding. Thanks to modern hormone testing and ultrasound exams, most ectopic pregnancies can now be detected before they cause serious problems.

At one time, ectopic pregnancy could be treated only with surgery, during which the fallopian tube was usually removed. Refined surgical techniques now make it possible to preserve the tube in many cases.

Medical (drug) treatment is now also available for the treatment of ectopic pregnancy. A drug called *methotrexate*, initially used to treat certain types of cancers, has been found to be extremely effective in selectively destroying ectopic pregnancy tissue and allowing it to be absorbed by the body. Methotrexate is a very safe drug, and may be given as either a single injection into a muscle or in several doses of shots or pills over the course of several days. Many early ectopic pregnancies can be treated in this way, often leaving the tube open. Methotrexate therapy cannot be used in all women, and may not work if an ectopic pregnancy is too far along.

Birth Defects

It is not clear whether birth defects are more common in babies conceived with ART. It may indeed be the case, however, because many women who pursue ART pregnancies are in their later reproductive years, when the risk of birth defects is naturally higher. Also, the frequency of multiple pregnancies is higher, and these gestations have an increased rate of birth defects. Studies have not shown that risks of birth defects arise directly from the use of ART. In addition, the types of birth defects in babies of ART pregnancies seem to be no different from those in other infants.

If Hopes Are Lost

Losing a pregnancy or a baby is always heartbreaking. For those who have struggled through infertility, it is an especially bitter loss. A miscarriage, however early it occurs in a pregnancy, is no less a loss than the death of a child, a parent, a friend, or a spouse. Society considers a period of grief natural following the death of a loved one, but sometimes people treat miscarriage more lightly. In many cases, miscarriage occurs before a couple has announced their pregnancy, and that means they may choose to grieve in private, with few, if any, friends and family members to support them. Whether you are alone in your grief or are dealing with a community that minimizes the impact of miscarriage on your emotions, you must know that the way to healing is to acknowledge your grief and work through it.

The Stages of Grief Everyone grieves a loss in his or her own way, but there are elements of grieving that are common to any experience of loss. In her widely acclaimed book, *On Death and Dying,* author Elisabeth Kübler-Ross identified five stages of grief. Many psychologists and other experts have since adapted her model to help explain the human response to loss. It is one way of understanding the powerful emotions of grief, and it may help you to find your own way through this process.

It is important to understand that these stages do not make up a timetable that you will follow in an orderly sequence. It is quite normal for you to go back and forth between steps, to "skip" steps, and to return to prior stages before finally coming to terms with your loss. But regardless of what path your own feelings take, it can help to understand that what is happening is normal and natural.

Shock and Numbness Upon first hearing emotionally painful news, you may react with shock and numbness. You may feel detached from your surroundings and strangely absent of feeling. You may think that you should feel upset but still feel nothing in response. You may find yourself thinking, for example, "This can't be true" or "There must be some mistake."

This is the mind's way of protecting itself from emotional pain. Just as the body's defenses rush in to heal a physical injury, so too does your mind work to block out emotional pain. Your initial lack of feeling does not mean that you do not care or are a cold, unfeeling person. As the reality of your loss sinks in, this feeling of numbness will pass.

Anger and Denial Once the initial shock wears off, you may begin to feel angry over what has happened. You may refuse to accept the reality of the painful event. Looking for something or someone to blame, you may lash out at the world, feeling that you are being punished unfairly. This is a normal reaction, as are envy, jealousy, and resentment toward other women and couples with healthy pregnancies and babies. It is also normal to feel anger at doctors or nurses, your partner, family and friends, and even God.

Feelings of anger might seem scary and unfamiliar, but they do not mean that you are crazy or a bad person. You do not need to worry about your mental health, either, unless you physically harm yourself or others, or become so hostile that you cannot function normally. Like other grief reactions, the anger is normal, and it will pass.

Guilt and Depression It is common for a person who has suffered a loss to begin to wonder whether she has somehow brought it on herself. From blaming others, you may eventually turn to blaming yourself. Anger and blaming that are turned inward—on yourself—lead to depression.

It is natural for you to want to find a reason for the loss of your baby. After all, so many other people have children, apparently without problems. Searching for a reason for what has happened may cause you to examine things that you have done in an effort to explain what has happened. You may begin to feel that you are at fault—for not living a sufficiently healthy lifestyle, for putting off having children, or for having had a prior abortion of an unwanted pregnancy.

There is no reason to think that any of these factors are the reason for your loss. You are not being punished, and you do not deserve your misfortune.

When anger and blame are turned inward, you may lose interest in things you once enjoyed. You may have trouble sleeping, sleep much more than usual, or lose your appetite. You may feel teary and cry often without knowing exactly why. Small things that normally would mean little to you may upset you greatly.

These are symptoms of depression. You are not going crazy, and there is a reason for your pain. Try to weather this emotional storm as best you can, reminding yourself that it will pass. Do things for yourself that bring you comfort—spend an afternoon with a good book, talking with a close friend or your partner, walking in the woods, or soaking in a hot tub. Take care of yourself by eating as well as you are able, and exercising if you can.

Bargaining At some point you may find yourself thinking that if you do certain things or behave in a certain way, things will improve and you will get what you desire. This may seem illogical, but it is natural during the grief process. Looked at in one way, it makes sense that if you feel that you are somehow responsible for your misfortune, then you should also have the power to reverse fate. Imagining that you somehow could gain control of your fate, however, is often a step on the path toward accepting your loss.

Acceptance and Resolution With time and patience, the pain of your loss will begin to subside. It will always be a part of you, but eventually it will no longer be the central focus of your life.

Guilt and the Fear of Forgetting The first time you realize that you haven't thought about your baby all day, you may feel guilty, as though you have forgotten him or her. But this is not so. Anyone who has suffered a great loss can tell you that you will never forget what has happened. The pain will recede over time, but traces of it will always remain. These traces are the memory of your loss, a part of your baby that you will always carry with you.

It is also quite normal to experience a rekindling of your grief on the anniversary of important dates. Such dates may be the day you learned you were pregnant, the day you first heard the baby's heartbeat, or the day you realized that your baby had died. On the anniversary of these days, you may have very intense emotions. You may even feel that you are going through the entire awful experience all over again. This is a common and normal reaction—so much so, in fact, that it has been named "anniversary reaction," or "shadow grief." It may help to anticipate the possibility that you may feel especially fragile on these days, so that you can make plans to alter your normal routine if you need to.

Remembering Your Loss At one time, parents who had lost a baby or who had miscarried were discouraged from thinking or talking about it. It was felt to be better for them if they did not openly acknowledge what had happened. Often, these couples would instead be urged to try for another pregnancy as soon as possible.

That thinking has very much changed. Many psychologists, doctors, and people who have had these experiences themselves say that the best way to come to terms with a loss is to confront it. In the hospital, parents who have lost a baby may be asked whether they would like to see and hold the baby, or to keep photos, a lock of hair, or the hospital name bracelet as a remembrance. If you've miscarried a pregnancy, you may wish to keep ultrasound pictures of the baby while it was still in the uterus.

To some people these things may seem morbid, but most parents find them enormously helpful. This is a way of saying good-bye—something you must do before you can heal, move on, and perhaps even begin to think about whether to try for another pregnancy.

Should You Try Again?

No one can answer this question for you. Only you know the limits of your resources. One thing is certain, though—it's best not to try to make this decision until you feel you have accepted your previous loss. Immediately throwing yourself into another round of infertility treatment is not considered the best way to heal from your grief and pain.

With time, some couples find that they have recovered enough to want to try treatment again. Others may come to a different conclusion: to live without children of their own. The next chapter presents some stories of people who confronted this decision. Their choices varied according to their own wishes, personalities, and lifestyles. None will fit your own story exactly, but some may inspire you to put your experiences into the context of others and to draw your own conclusions.

8

When Treatment Fails: Facing Infertility and Finding Alternatives

For many couples facing infertility—about 35 percent of those who try treatment—there comes a time when they wonder whether it is time to stop. People reach this point at different times in the treatment process. Some wait until they feel they've tried every possible method of treatment; others draw the line at an earlier point, not wishing to or not able to expend more time, energy, and money.

Although thousands of people each year are able to conceive through assisted reproductive techniques (ART), many more run up against the limitations of even these advanced medical technologies. When treatment does not succeed, many people choose other pathways to parenthood, such as surrogacy or adoption. Others come to terms with life without children and put their energies into other meaningful endeavors.

This chapter discusses the decision to end infertility treatment and presents other options for becoming parents or living without children.

How Do You Know When It Is Time to Stop Trying?

After all you have been through—after the seemingly endless decision making, the unsuccessful intrauterine inseminations and/or in vitro fertilization cycles, the waiting and hoping, and the repeated disappointments—how do you know when it is time to stop infertility treatment? The decision comes more easily for some people than for others.

Your attitudes concerning childbearing and parenting may have changed since you started seeking fertility treatment. At any rate, the day-to-day demands may have made it difficult to concentrate on them. Now is the time to reevaluate your feelings on the subject.

Where Do You Stand?

People differ not only in their reasons for wanting children, but also in the degree to which they want them. Consider the range of feelings reflected in the following responses to the question, "How do you feel about having children?"

- From a 37-year-old man: "I have always known, from the time I was very young, that I wanted to be a father someday. Last year my wife and I had our first child, and I have never been so happy. I want a big family with four or five children. I dream about the day that our house will be filled with the happy chaos of children."

- From a 41-year-old woman: "I guess I always thought that I would have children someday. Obviously, it's what my parents had done, and it's also what my brothers and sisters have done, so I thought I would, too. But I've been married for eight years now, and my husband and I have been very happy. I realize that maybe this is not something I have to do—or want to do—after all."

- From a 26-year-old man: "I never really thought about having children—until my best friend from childhood became a father. There have been other children in my life, but when I first saw his baby, it was like a switch went off somewhere, and I knew I wanted kids of my own."

- From a 57-year-old woman: "I have never felt the desire for a child that some women have. Both my husband of thirty-three years and I love children very much, and we have many in our lives—those of close friends, of siblings, and of neighbors. But we are also very happy with our life the way it is. I know I may not be like most women my age, but I am going on 60 now, and I have never for a moment regretted being child-free."

Where do your own feelings now lie? Have they changed from what they were, say, a year ago? Five years? Ten? Often, people take for granted ideas they have held since youth, when actually their feelings may have changed over the years. For some, the experience of infertility treatment serves to sharpen their desire for a child. For others, it causes them to pause and question that desire.

Childbearing: What It Means to You

As more women in the baby-boom generation have children of their own, the rate of births in the United States is rising. About three million

American women gave birth in 1976. Nearly twenty years later, in 1994, that number had risen to almost four million.

With this so-called "baby boomlet," both women and men are bombarded with the message that having children is the *expected* thing to do. Many infertile couples have a heightened sensitivity to these societal messages. "Family-oriented" (usually meaning "child-oriented") stores, restaurants, theme parks, TV stations, and even websites all serve as reminders to the childless couple that they are not considered by society's mainstream to be a "real" family.

It can be difficult, in the face of all this, to sort out your own feelings about childbearing. The idea of having children means different things to different people. Many people go through their lives without giving the matter much thought. But people who have struggled with the pain of infertility have not had the luxury of taking for granted their ability to reproduce. If you are like most of these women and men, your ordeal has forced you to think a great deal about childbearing.

Now, at the juncture where treatment has failed you, it is likely that you will be forced to review these issues once again. In order to decide where to go from here, you will need to reexamine the meaning that childbearing has for you. It may help to revisit Chapters 3 and 6 to sort out your feelings and identify the underlying beliefs and values that are at the center of this decision.

What are these core values and beliefs? What are the most important aspects to you of having a child? The list that follows includes some of the reasons why many couples say they want to have children. There may be others that you can add to the list. All of these factors go along with having children, but some will be more or less important to you, depending on your own values.

- Having the firsthand experience of carrying a pregnancy to term

- Seeing/being present at the baby's birth

- Having a child who is the biological offspring of you and/or your partner

- The family connection: Giving a grandchild to your parents and in-laws, a niece or nephew to your siblings, and/or a cousin to your nieces and nephews

- Remaining in, or rejoining, your social circle of friends who have children

- Passing on your own and your family's physical traits, mental attitudes, and moral beliefs and values

- Knowing that something of yourself will be passed on to the next generation and will remain after you are gone

- Having someone in your life to whom you are unconditionally devoted

- Being responsible for a new life and for raising a child to adulthood

Talking It Over

It may help you and your partner to identify the degree of importance that each of the factors described above has for you. As an exercise, you could try rating each one—for example, as most, somewhat, and least important—and comparing your answers. This can provide fuel for discussion and reflection. You may come to realize that having children is just as important as ever, or that there are other ways of meeting the goals that are most important to you.

As you begin to consider stopping fertility treatment, remember that your own feelings may differ from your partner's. Start by simply listening to what the other person has to say without reacting negatively or defending your own thoughts. If you are at wildly different points on the question of whether to stop or continue, it may be wise to seek the help of a professional counselor or therapist. (You may want to review Chapter 6 for suggestions on improving communication.)

Deciding to Stop

Many couples decide to stop infertility treatment before trying all that the field has to offer. Some report that one of the greatest difficulties they faced in choosing to stop was guilt, especially in the face of pressure from family, friends, and society to become parents.

Here is something to remember: Just because a treatment option can be tried does not mean that it should be tried. Do not let others' pressures and expectations push you into a long, expensive, and often frustrating course of treatment that you do not really want. You do have other options. Some are discussed later in this chapter.

When the moment of truth arrives, and you and your partner have carefully considered your current feelings on the subject of childbearing, the following practical suggestions may be useful.

- Consider setting a time frame. Having a schedule can be helpful, even if you decide to change it later. You could choose, for example, to continue treatment for another year, or until your or your partner's next birthday. When the date arrives, you can always decide to continue. But at least it will make you and your partner pause and talk it over first.

- Think about taking a "vacation" from treatment. See how you feel if you stop for awhile. You might feel a sense of relief and become clearer about what steps you want to take next. You might become sad and depressed, too. Allow yourself to have these feelings. Eventually your decision will become clearer to you.

- Professional counseling may also help you make this decision.

• Talk to others who have decided to end treatment. Ask them how they reached this decision and how they feel about it now. Learning about the experiences and feelings of the couples described in the rest of this chapter may help you to sort out your own.

Alternatives to Pregnancy

There are many roads to parenthood. Having a child through a pregnancy of your own is one. Adoption and foster parenting are others. However, especially for couples who have dreamed of experiencing pregnancy and the birth of their child, adoption and foster parenting may seem like unsatisfactory second choices. Many couples fear that they will never be able to love a child who is not their own biological offspring.

It is true that not everyone is suited for adoption or foster parenting, and the decision to choose either must not be taken lightly. However, if reservations can be wholeheartedly overcome, both options can provide a family for the adoptive parents as well as for the adopted children.

Adoption

Once you begin to consider adoption, you are likely to find that in many ways a family with adopted children is the same as any other family. As your children grow, you will express love, have disputes, and make compromises in your daily lives. And as adoptive parents, you will have legal responsibility for the welfare of your children.

But you will also come face-to-face with several issues that occur only in adoptive families. When and how will you discuss with your child the fact that he or she is adopted? How will you handle your child's feelings about not knowing his or her biological parents? Feelings of rejection are not uncommon among children who are adopted, even when their adoptive parents are showering acceptance and love on them. How will you feel if, for a time, your love is not enough for your child? And how will you all cope with insensitive people who assume that adoption is a second-best alternative?

There are many support groups that can provide you with information and counseling to deal with these issues. You are likely to encounter these organizations during the adoption process. Take advantage of all the support groups and preparatory classes you can prior to choosing adoption.

LINDA AND BOB: LOOKING FARTHER AFIELD

"After five years of infertility treatment," says Linda, a 42-year-old ex-actress, "we were both in despair."

Bob, 46, agrees. "It almost broke us apart."

"The months of getting our hopes up, only to have them dashed," says Linda, "really took a toll on us. I had two miscarriages, which was incredibly hard. I started feeling guilty because years ago, when I was a very young woman, I'd had an abortion. It felt like I was being punished."

"It was especially hard because we never found the reason for the infertility," says her husband of sixteen years. "After awhile, we were losing all perspective. One day we just looked at each other and said, 'Why are we going through this? There are other ways to have children.' We decided we would start looking beyond what infertility medicine had to offer."

As Linda and Bob talk about their years of trying to become parents, their two children, Hannah and Jamie, play at their feet. Hannah was adopted in 1989; Jamie's adoption followed in 1992. Both children were found through their adoptive parents' connection with a private-practice obstetrician in another state. An old friend of Linda's, the physician had been called in the middle of an April night to attend a very frightened 16-year-old who was in labor. The girl lived with her stepfather and sister and had managed to keep her pregnancy a secret until labor started. Her sister drove her to the hospital.

Eighteen hours later, the girl, barely out of childhood herself, gave birth to a healthy baby girl. Talking to her afterward, the doctor found her adamant about giving up her new daughter for adoption.

"With our daughter, the adoption process was very iffy," says Linda. "Because the birth mother was underage, there were complications about her consenting to the adoption. And the boy she named as the father refused to acknowledge paternity."

"That left us in a bind for months," says Bob. "The fact that she named the father—as opposed to just saying the father was 'unknown'—meant that his consent was needed. And he was denying paternity, so it was a catch-22 situation. At one point we were really scared we would lose her—this was after Hannah had already been placed with us. Finally, the birth mother agreed to list the father as 'unknown.' A few months later, Hannah became our daughter legally."

Jamie's story was altogether different from Hannah's. When he was born, his birth mother had four children and, single and with limited means, could not raise a fifth. Not wanting an abortion, she decided to give up her fifth child to Linda and Bob.

Although the adoption went smoothly, there were other problems. Linda and Bob are both white. Jamie is African American. "We were so starry-eyed about Jamie," says Linda, "that I'm afraid we were a bit naive about the challenges of being an interracial family. As a family, we were OK. Hannah adored her little brother. And both kids knew from day 1 that they were adopted. But one day Hannah came home from kindergarten crying because some other kid had said that Jamie couldn't be her 'real brother' because he is black and she is white.

"I realized that, in our idealism, we hadn't fully prepared Hannah for the fact that other people could make life hard for her and her brother."

Linda and Bob handled this particular situation by explaining to Hannah that Jamie was just as much her brother as if he had been born by the same mother. "I was really proud of how she handled that kid the next day," says Bob. "Her teacher said that Hannah marched in there and told that kid that it didn't matter what color her brother was. He was her brother because she loved him."

After the years of pain they had experienced over their childlessness, Linda and Bob now say that they are glad they made the choice they did. "It's not easy being a parent," says Linda. "We have a lot of the same problems that any family has," agrees Bob, "and being an adoptive interracial family creates other problems as well. But the bottom line is, we love our children. We've been able to give them something that their birth parents could not."

"And they," agrees Linda, "have given us something that we couldn't give ourselves. In the end, that's what it means to be a family."

Types of Adoption At first, the choices open to you in adopting a child may seem overwhelming. There are many factors to consider, and you will need to sort through them all to decide what sort of adoption process and what kind of child you are looking for.

Open and Closed Adoptions At one time, "closed" adoption was the standard. In this arrangement, the adoptive parents and the child's biological parents would have little or no knowledge of each other and usually would never meet. The adoptive parents had only minimal information about the birth parents. Every effort was made to keep the adoption records closed and unavailable to anyone—including the adoptive parents, the birth parents, and the adopted child. Most agencies believed that the complete separation of the adoptive parents from the birth parents was necessary for the adoptive family to have a normal parenting experience.

In recent years, "open" adoption has gained widespread acceptance. There are many variations on the concept of open adoption, but it generally involves at least a limited exchange of information and contact between the birth mother (and sometimes the birth father) and the adoptive parent or parents. Some agencies even allow the birth parents to choose the adoptive parents. These agencies give the birth parents anonymous information on selected couples that meet their criteria. An open adoption often involves a face-to-face meeting between the adoptive and birth parents.

Infant and Older Child Adoptions Both infants and older children can be adopted. Although parents of any age may adopt infants through a private arrangement (see page 197), most agencies are reluctant to place infants with parents over 35 or 40 years of age. If an older child is being adopted, agencies usually require that there be at least a fifteen-year age difference between parent and child. Couples older than 40 may be encouraged to adopt older children.

Interracial Adoptions It is not uncommon for parents of one race to adopt children of another. If you decide on interracial adoption, you should be aware of the issues most interracial families face over the years following the adoption. Most agencies provide counseling for parents considering this option. A number of organizations exist to offer support, education, and counseling for both parents and children in interracial adoptions (see Appendix).

Adoption of Children with Special Needs In most cases, the term *special needs* refers to problems with a child's emotional or physical health. Adopting a child with mental or physical differences can be demanding, and also enormously rewarding. It requires a strong commitment. Parents who choose this option usually do so with a full understanding of the demands of raising a child with differences. Most find the experience gratifying, and some adopt a second child with special needs.

   If you are interested in adopting a child with special needs, talk with parents who have done so—those who have been successful and those who encountered problems they could not surmount. It is important to have a realistic understanding of your child's needs and of the resources you will need to meet them.

Ways to Adopt There are several routes by which you may adopt a child. Laws and regulations vary by state, and it is best to get acquainted with them before exploring this option. The following is an overview of the ways in which children may be legally adopted in the United States.

Agency Adoption Adoption agencies may be either public or private. Their mission is to find appropriate homes for children who need them. Public agencies are usually run through the state and are free. Private agencies generally charge fees, sometimes according to a sliding scale, which may include the expenses of the birth parent(s).

Agencies tend to have different policies and procedures, so you will need to find one whose style you feel comfortable with. Places to find adoption agencies include your local yellow pages, as well as your state's department of social services. The best way to choose an agency is by word-of-mouth endorsement. If you know someone who has adopted through an agency, by all means ask them about their experience. There also may be adoptive-parent support groups in your area. Learning how others went about adopting a child, and with what agency, can save you a lot of time and emotional energy.

Some agencies will walk you through each step in the process. Others expect you to take the initiative in gathering information. When you investigate agencies in your area, ask about their policies concerning matters such as confidentiality, and find out whether they specialize in a certain type of adoption (older children, international adoptions, or interracial adoptions). It is also a good idea to talk with individuals or couples who have used a particular agency. If you check with your state's agency licensing department, you can find out whether any complaints have been lodged against an adoption agency.

There are many books on adoption that can help get you started and guide you through this process. Some are listed in the Appendix.

Private Adoption A *private adoption,* or *independent adoption,* is an arrangement made between people who contact each other directly, without an agency acting as a go-between. If you begin the process of arranging for a private adoption, it is imperative that you first learn about the laws in your state. If the birth parent is living in another state, you will need to research the laws in that state, too. To go into an adoption without knowing all the legal issues is to invite serious problems in the future.

Couples wishing to adopt privately may make their wishes known through classified ads, word-of-mouth, contacts with physicians or lawyers, or computer database listings. Not all states allow all of these methods, though, so be sure you know what is legal before trying them.

There is almost always the risk, in private adoptions, that the birth mother or birth father will decide at some point not to go through with the process. The law requires a waiting period in a private adoption, so that even after the child is placed with the adoptive parents, there is a set period of time during which either birth parent can reverse his or her decision to give up the child. Obviously, the emotional consequences of this can be terrible for the adoptive parents.

Identified adoption is a type of private adoption in which the adoptive parents and the birth parents find each other independently. They then go to an adoption agency or attorney to complete the adoption process.

It is best to have a thorough knowledge of the laws in your state and of the possible outcomes of any private adoption arrangement. Consulting an adoption attorney (see Appendix) may prove helpful in this regard.

International Adoption Each year, thousands of children enter adoptive American homes from other countries. These adoptions are arranged independently by adoptive parents applying directly to orphanages, maternity homes, or lawyers in other countries. Most, however, are arranged through the many agencies in the United States that specialize in international adoptions.

Some United States agencies have adoption programs for several different countries. Because each country has its own laws and regulations governing adoption, the agency and the adoptive parents must meet the requirements established by the country in which the child is born. The paperwork involved in an international adoption can be formidable—not only are you satisfying the laws of the child's birth country, but also those of your own state and the United States government. This can be extremely time-consuming and expensive.

Babies adopted from abroad may have health problems that, although not permanent or long-term, will need special attention. Although this route of adoption can be difficult, there are thousands of children in foreign countries who need a stable home and loving parents.

Foster Parenting

Becoming a foster parent is different from adopting a child in that, although you may have physical custody of the child, the arrangement is usually temporary. Foster care involves the placement of children with single adults or couples. In most instances, a social service agency seeks foster parents for children whose own parents have not cared for them appropriately. The arrangement is usually considered temporary in the hope that the family of origin can eventually be reunited.

Each state has its own child protective services agencies that usually recruit and train foster parents. To become a foster parent, you must be at least 21 years of age. States have upper age limits as well, usually age 65. The prospective parent(s) must demonstrate financial stability. Other requirements vary according to the state in which you live. Many foster parents have children of their own in addition to their foster children.

If you are thinking of becoming foster parents, you will probably be required to attend a series of workshops or orientation sessions. You will be screened to ensure that you meet the state's requirements, and you will then submit an application to foster a child. At some point, a home visit will be made by agency staff to discuss your personal and family history, lifestyle, experiences with child care, and the types of children you feel you would be able to care for.

The children who need foster care vary in age, race, and family background. Many have been neglected, abused, or abandoned. They range in age from infancy to older teens. They may be only children, or part of a group of siblings who are in foster care. As a foster parent, you are responsible for your child's daily care and nurturing, for making sure their needs are met in school and in the community, and for keeping in touch with caseworkers to apprise them of any problems. If it is established that reuniting the child with his or her family of origin is a realistic goal, you will be expected to help the child work toward that.

Obviously, foster parenting requires a different kind of commitment than adopting a child. Most foster parents must be prepared to cope with a child who has a troubled background and then to allow the child to move on, if circumstances permit. Thousands of single people and couples have discovered that foster parenting is a richly rewarding experience. It is a way to have children become an integral part of their lives while providing children with a safe, stable, and loving family atmosphere.

Alternatives to Parenthood: Child-Free Living

> *"A successful marriage depends on the presence of children . . .*
> *most marriages are far more likely to be permanent and happy if*
> *children come soon to complete the marriage bond."*
>
> From a family medical reference book, 1941[2]

Many women and men today would strongly challenge the attitude reflected in this quote. Still, some part of this assumption about marriage remains in our society. It can contribute to the isolation that many infertile couples feel. It can also add to your confusion if you, like many people, are grappling with the question of whether to have children of your own.

Choosing to live without children is an option that is always open to you, regardless of where you are in your infertility treatment. If you are like some infertile couples, you may know right away that this choice is not for you. If you are like others, though, you may have begun to question your desire for children, and you may want to consider living child-free.

---

[2] *Modern Home Medical Adviser.* Edited by Morris Fishbein, M.D. New York: P. F. Collier & Son, 1941.

Choosing to Be Child-Free

Many couples who have experienced infertility have opted to end treatment and embrace a lifestyle that does not include having children in the home. The reasons for their choice are as many and varied as the individuals themselves. Some infertile couples find that, upon reaching a certain point in treatment, they do not wish to proceed with more technically advanced procedures. Others, after trying all the medical options available, have reevaluated their priorities and have chosen to direct their energies to other pursuits.

You and your partner may have reasons of your own for considering this choice. You may just want to take a break from treatment. You can almost always resume treatment at a later date, if you wish. In the meantime, you may want to see what the child-free lifestyle is like when it is something that you have consciously chosen.

The story presented here is of Amy and Alex, who decided to end infertility treatment after several failed IUI attempts. Their experience is just one example of how a couple might arrive at this decision, and how their lives might change as a result.

❧ AMY AND ALEX: CHANGING COURSE

When Amy and Alex decided to have a child after three years of marriage, their parents were overjoyed. "My mom and dad were especially elated because I am an only child," says Amy. "And Alex's mother was overjoyed because he is the only one of her children who is married."

As his wife speaks, Alex looks out across a beautiful pasture, where horses are grazing. They live on a small, rural farm where they board eight horses and keep two of their own. "If we hadn't changed course," he says quietly, "we wouldn't be here right now."

Amy and Alex's hopes of raising children were dashed when Amy did not become pregnant. After more than a year of trying, the couple consulted a fertility expert. Tests showed the most likely cause of their infertility to be Alex's low sperm count. The couple tried three times without success to conceive through IUI using Alex's sperm.

"At that point," says Alex, "we realized we were at a crossroads. The next step would probably be IVF. But we weren't so sure we wanted to take that path."

"It was really hard because of all the family pressure," says Amy. "But we talked it over and decided to give ourselves a few months off from treatment to see how we felt. I guess what happened was, the 'waiting' period never really ended. We felt closer than ever, and we were really enjoying our lives—especially without the pressure we'd been under during the treatment. Then this place came up for sale."

Amy is an avid equestrienne, and Alex was raised on a horse farm. The two say they have always had only two dreams—to have children and to own a horse farm.

"We realized that we could afford to buy the property," says Alex, "if we used the money we'd saved for infertility treatment." As he speaks, his red hair is ruffled by a breeze, and his 6-foot frame rests against a wooden fence. "And I guess we both just knew in our hearts that we wanted this— and that we didn't want to pursue the treatment any longer."

Amy agrees. "It was the best decision we ever made. It was the right time to stop for us. I know we could have tried more options, but we both really felt that we had done as much as we could."

"The hardest part of deciding not to have children wasn't our own feelings—it was the reactions of our family,"

Alex says. "We knew our parents were hurt, and that was really hard. But it wasn't enough of a reason to continue treatment when we didn't want to continue. And in the end, I think they have come to understand."

Now, Amy and Alex have children around them much of the time. Alex teaches riding to school-aged children, and Amy volunteers for the local Pony Club, whose members are 16 and under. In addition, several of their boarders have children who come with them to the barn, and the couple has grown close to them.

"We are having the time of our lives," says Amy.

Attitudes Toward Childbearing in the Late Twentieth Century

Many attitudes about women's roles in both the home and the workplace have changed over the past several decades. The idealized family portrayed on popular TV shows in the 1950s was headed by a father who went to work and a mother who stayed home. Today, half a century later, a new ideal is often expressed by the catch-phrase "having it all." For women, this usually means juggling the responsibilities of having a career in addition to—not instead of—fulfilling the "traditional" roles of wife and mother. But despite American women's entry into the workforce—and the resultant delaying of childbearing by many—often an assumption remains that they will, eventually, want to become mothers.

Because of assumptions such as these, many women and men—especially those without children—feel that societal expectations and attitudes about childbearing have not really changed. Pressure from well-meaning friends, family, and coworkers can make it hard to feel entirely free to make the conscious choice to remain childless. Many couples who are childless by choice say that the reactions of family, friends, and coworkers to their decision have ranged from confusion to downright hostility.

But increasing numbers of couples are making this choice, and at increasingly younger ages. One 25-year-old woman who recently had a tubal ligation says that, although her family has been supportive, others do not always understand her decision. "A lot of people seem to feel that my choice to be childless is foolish because it is such a permanent decision. They think that I can't possibly know at my age what I really will want in the years to come. But I remind them that having a child at my age would also be an irreversible, life-altering decision."

Your Own Journey

Whatever choices you make in the course of infertility treatment, they must be the right ones for you. This book has suggested some ways in which to sort out your feelings, especially during times of stress and pressure from outside. If you are pursuing treatment with a partner, it is especially important to keep the lines of communication open. When you hit bumps in the road that you cannot seem to get past by yourselves, counseling is a helpful option.

When you are in the midst of a journey, it can be hard to know where you will end up. At times, the future may seem frightening and overwhelming. At such times it will be important to focus your view on those things that you need to consider right then. As you progress toward your goals—which themselves may change in the course of your trip—it is good to know that your steps are ones that you have chosen freely.

Appendix

General Information on Infertility and Women's Health

The **American College of Obstetricians and Gynecologists (ACOG)** is a private, voluntary, nonprofit, membership organization for professionals who provide health care for women. ACOG promotes patient education and involvement in medical care and works to increase awareness among its members and the public of the changing issues facing women's health care.

ACOG
409 12th Street, SW
Suite 409
P.O. Box 96920
Washington, DC 20024
Phone: (202) 683-5757
E-mail: mgraves@acog.org
Website: http://www.acog.org

The **American Society for Reproductive Medicine (ASRM)** is a voluntary nonprofit organization devoted to advancing knowledge and expertise in reproductive medicine and biology. ASRM fosters patient understanding and involvement in reproductive medicine through the Patient Information Series of booklets and a variety of other information sources that are free to the public.

The American Society for Reproductive Medicine (ASRM)
Patient Information Department
1209 Montgomery Highway
Birmingham, AL 35216-2809
Phone: (205) 978-5000
Website: http://www.asrm.org/
Email: asrm@asrm.com

The **Center for Male Reproductive Medicine and Microsurgery** at Cornell University Medical College provides state-of-the-art, compassionate care for the infertile male. The center carries out basic and clinical research in male reproduction, trains residents and fellows, and provides educational programs for professionals as well as the public.

Center for Male Reproductive Medicine and Microsurgery
The New York Hospital-Cornell Medical Center
525 East 68th Street
New York, NY 10021
Phone: (212) 746-8153
Website: http://www.maleinfertility.org

The **Endometriosis Association** is a nonprofit, self-help organization founded by women for women. It is dedicated to providing information and support to women and girls with endometriosis, educating the public as well as the medical community about the disease, and conducting and promoting research related to endometriosis.

Endometriosis Association
8585 North 76th Place
Milwaukee, WI 53223
Phone: (414) 355-2200; toll free 1-800-992-3636
E-mail: endo@endometriosisassn.org
Website: http://www.EndometriosisAssn.org

The **Ferre Institute** ("Ferre"—from the Latin word "to bear") is a nonprofit organization devoted to educational services that focus on infertility, reproductive health, and family-building. Among its services is education for infertile couples on the availability of resources for treatment and resolution, including adoption. The institute sponsors symposia and workshops; maintains a resource library; and produces a newsletter, videotapes, and brochures (*Reviewing Your Options*; *Your Role on the Medical Team*; *Helping Others Understand*; *Coping with the Holidays*; *Miscarriage: Surviving Pregnancy Loss*; *We Can't Have a Baby Either*; *Answering Your Questions About Infertility*).

Ferre Institute, Inc.
258 Genesee Street
Suite 302
Utica, NY 13502
Phone: (315) 724-4348
E-mail: FerreInf@aol.com
Website: http://members.aol.com/ferreinf/ferre.html

In addition, a reference (non-circulating) library is available to professionals and the public at the institute's Genetic Counseling Program:

124 Front Street
Binghamton, NY 13905
Phone: (607) 724-3408
Fax: (607) 724-8290

The **InterNational Council on Infertility Information Dissemination** (INCIID—pronounced "inside") is a nonprofit organization committed to providing the most current information regarding the diagnosis, treatment, and prevention of infertility and pregnancy loss. This information is disseminated primarily via the latest information technologies, such as the Internet and commercial online services. INCIID's purpose is entirely educational.

INCIID
P.O. Box 6836
Arlington, VA 22206
Phone: (703) 379-9178
E-mail: INCIIDinfo@inciid.org
Website: http://www.inciid.org

Infertility Support

RESOLVE is a national nonprofit agency that serves as a counseling, referral, advocacy, and support system for infertile couples and offers education and assistance to associated professionals and patients. RESOLVE has chapters in more than fifty-five cities across the country, where support groups and programs are conducted. The national office publishes a quarterly newsletter for a $45 membership fee and provides numerous fact sheets and reprints on the issues of infertility and related subjects. For a general mailing, send a self-addressed, stamped, business-size envelope to the address below.

> RESOLVE
> 1310 Broadway
> Somerville, MA 02144-1779
> Business office phone: (617) 623-1156
> National HelpLine: (617) 623-0744
> E-mail: resolveinc@aol.com
> Website: http://www.resolve.org

Infertility Alternatives

Adoptive Families of America (AFA) is a private, nonprofit membership organization of families and individuals. AFA provides problem-solving assistance and information about the challenges of adoption to members of adoptive and prospective adoptive families. AFA seeks to create opportunities for successful adoptive placement and promotes the health and welfare of children without permanent families.

AFA
2309 Como Avenue
St. Paul, MN 55108
Phone: (612) 645-9955; toll-free 1-800-372-3300
E-mail: info@AdoptiveFam.org
Website: http://www.adoptivefam.org

AFA has assembled a guide for people considering adoption as a way to build their families and as a solution for wanting children. Call 1-800-372-3300 for a copy of *The Open Adoption Experience: A Complete Guide for Adoptive and Birth Families, From Making the Decision Through the Child's Growing Years*, by Lois Ruskai Melina and Sharon Kaplan Roszia, HarperCollins, 1993, $16.00.

The **American Academy of Adoption Attorneys (AAAA)** is a national association of approximately 300 attorneys who practice, or have otherwise distinguished themselves, in the field of adoption law. The academy's work includes promoting the reform of adoption laws and encouraging ethical adoption practices.

AAAA
1300 19th Street, NW
Suite 400
Washington, DC 20036
Phone: (202) 832-2222
Website: http://www.adoptionattorneys.org

American Foster Care Resources is a nonprofit organization providing information and educational resources for and about family foster care.

American Foster Care Resources
P.O. Box 271
King George, VA 22485
Phone: (540) 775-7410 (Monday through Friday, 9:00 A.M. to 4:00 P.M. Eastern time)
E-mail: afcr@lava.net
Website: http://www.lava.net/~afcr

The **North American Council on Adoptable Children** (**NACAC**) links adoptive parents' groups in the United States and Canada. They are involved in child and family advocacy as well as public information and education.

NACAC
970 Raymond Street
Suite 106
St. Paul, MN 55114-1149
Phone: (612) 644-3036
Website: http://www.cyfc.umn.edu/adoptinfo/nacac.html

Adoption and Foster Care Information on the World Wide Web

Adoption: Assistance, Information, Support
http://www.adopting.org

Foster Parents CARE
http://www.fostercare.org

Books on Adoption and Foster Care

Adopting After Infertility, by Patricia Irwin Johnson. Perspective Press, 1992.

The Adoption Resource Book, by Lois Gilman. HarperCollins, 1992.

The Long-Awaited Stork: A Guide to Parenting After Infertility, by Ellen Sarasohn Glazer. Jossey-Bass Publishers, 1998.

Glossary

Acrosomal cap A microscopic structure that partially covers the tip of the sperm cell and that produces chemicals that help to break down the outer membrane of an ovum at fertilization.

Adhesions Scar tissue that sometimes forms inside the abdomen after abdominal or pelvic surgery or infection.

Adrenal glands Two hormone-producing glands located on either side of the abdomen, just above the kidneys.

Amenorrhea The absence of menstrual periods, usually caused by hormone imbalance.

Andrology The branch of medicine concerned with the diagnosis and treatment of disorders of the male reproductive system.

Anejaculation A condition in which no semen is expelled from the penis during sexual arousal.

Antigens Proteins that are carried on the surface of a foreign substance in the body and that trigger an immune response.

Assisted hatching A technique performed after fertilization with IVF to improve the implantation of the embryo in the uterine wall.

Assisted reproductive techniques (ART) Procedures in which pregnancy is attempted through the use of external means, such as in vitro fertilization (IVF) or gamete intrafallopian transfer (GIFT).

Azoospermia A cause of male infertility in which no sperm are present in the semen.

Basal body temperature A woman's baseline temperature, which rises by about one-third to one-half of a degree Fahrenheit at ovulation; this increase can be charted to determine when ovulation is occurring and pregnancy is most likely.

Beta-thalassemia An inherited blood disorder found especially among peoples of Mediterranean descent and that is linked to infertility in women.

Bioethics The study of moral issues in medical treatment and research.

Biological father The man whose sperm fertilized the ovum from which a child developed and who is therefore genetically related to that child.

Biological mother The woman from whose ovum a child developed and who is therefore genetically related to that child.

Birth canal The passageway leading from the uterus, through the cervix and vagina, and to the outside of a woman's body, through which the fetus passes during labor and delivery.

Blastula The zygote at about four to five days after fertilization, when it has developed into a two-layered structure of cells surrounding a fluid-filled cavity.

Breech presentation A situation in which the baby is positioned feet down inside the uterus at the time of labor and delivery.

Bromocriptine A medication used to treat excess prolactin production by restoring that hormone to normal levels.

Bulbourethral glands Two small glands that are located just below a man's prostate gland and that produce fluid to aid the movement of sperm.

Cabergoline A medication used to treat excess prolactin levels.

Cervix The lower, narrow end, or neck, of the uterus, which opens into the vagina.

Chlamydia A sexually transmitted disease caused by the microorganism *Chlamydia trachomatis*, which if left untreated in a woman may cause pelvic inflammatory disease (PID).

Chromosomes Submicroscopic structures that are located inside each cell in the body and carry the genes that determine a person's physical traits.

Cilia Specialized hairlike cells inside a woman's fallopian tubes that aid the movement of sperm.

Cirrhosis of the liver A disease, caused by alcoholism or infections such as hepatitis, in which the cells of the liver are damaged so that they cannot properly filter waste products from the blood.

Clomiphene citrate A medication used to stimulate ovulation.

Controlled ovarian hyperstimulation (COH) A technique to enhance ovulation by administering hormone medications that stimulate the development of multiple eggs for use in an insemination or IVF technique.

Cowper's glands See bulbourethral glands.

Crohn's disease See inflammatory bowel disease.

Cryopreservation A special freezing technique used to preserve embryos and oocytes for future use in an ART procedure.

Cryptorchidism A cause of male infertility in which one or both testes have not descended into the scrotum after the first year of life.

Delayed ejaculation A condition in which ejaculation occurs only after a very prolonged period after sexual arousal.

Dexamethasone An anti-inflammatory drug sometimes used with ovulation-inducing drugs.

Diethylstilbestrol (DES) A medication given to some women in the 1950s to prevent miscarriage, now known to be harmful to the growing fetus.

Dilation and curettage (D&C) A procedure in which the cervix is gradually widened and the lining of the uterus is gently removed by scraping or suction.

Ductus deferens See vas deferens.

Dyspareunia A condition in women in which intercourse is uncomfortable or painful due to a medical problem or, rarely, emotional issues.

Early menopause See premature ovarian failure.

Eclampsia A condition in which seizures occur in a woman with high blood pressure during pregnancy.

Ectopic pregnancy A pregnancy in which the fertilized ovum implants in a location other than inside the uterus, usually in a fallopian tube.

Electroejaculation A method of sperm collection in which a low electrical voltage is conducted into the nerves that stimulate ejaculation.

Embryo The fertilized ovum after it has begun the process of cell division.

Endocrine glands Glands that secrete hormones to regulate various bodily functions.

Endometriosis A condition in which tissue resembling that lining the inside of a woman's uterus is found elsewhere in the abdomen.

Endometrium The tissue lining the inside of a woman's uterus, in which a fertilized egg implants at conception.

Epididymis A structure that covers part of each testis and is the storage place for mature sperm cells.

Epididymitis Infection of the epididymis (the structure covering part of the testis in which sperm are stored), often caused by an untreated sexually transmitted disease.

Estrogen A hormone that is produced in a woman's ovaries and plays a role in regulating ovulation.

Fallopian tubes The two narrow, tubelike structures located on either side of a woman's uterus in the lower abdomen, extending to an ovary on each side.

Fee-for-service A traditional form of health insurance in which the insured person pays a monthly fee to the insurance company and pays a predetermined percentage of providers' fees as they are incurred.

Fetal growth retardation See intrauterine growth retardation.

Fetal reduction See selective reduction.

Fetus In medical terms, a fertilized egg becomes a *fetus* at about the end of the seventh week of pregnancy, after major structures (head, torso, limbs, etc.) have formed. In general parlance, the term *fetus* is often used to describe a baby-to-be at any point from fertilization until birth.

Fibroids See uterine fibroids.

Fimbriae The fingerlike projections at the end of each fallopian tube near the ovary.

Follicle A structure containing the egg that is produced at ovulation.

Follicle-stimulating hormone (FSH) A hormone that is secreted by the pituitary and that, with luteinizing hormone (LH), signals the testes to produce sperm in males and the ovaries to produce ova in females.

Galactorrhea A disorder, caused by a hormone imbalance, in which breast milk is produced by a woman who is not pregnant or breast-feeding.

Gamete intrafallopian transfer (GIFT) A variation of IVF in which unfertilized eggs and sperm are placed together in the woman's fallopian tubes, fertilization taking place in the tube instead of in a laboratory dish.

Germ cells The sex cells—the ova in females and the sperm cells in males—each of which contains twenty-three chromosomes and that combine at fertilization to make up the fetus's full complement of forty-six chromosomes.

Gestational mother In a surrogacy arrangement, the woman who carries a pregnancy to term and delivers a baby, which may or may not be genetically related to her (*see biological mother*).

Gonadotropin-releasing hormone (GnRH) A hormone that is secreted by the hypothalamus and that signals the pituitary to release other sex hormones.

Gonadotropins Hormones that are injected to stimulate ovulation.

Gonorrhea A sexually transmitted disease caused by the microorganism *Neisseria gonorrhoeae*, which if left untreated in a woman may cause pelvic inflammatory disease (PID).

Health maintenance organization (HMO) A type of managed care health insurance plan in which the insured person chooses a primary care physician from a preapproved list of providers.

Human chorionic gonadotropin (hCG) A hormone normally produced by the growing fetus and that, in synthetic form, may be used to stimulate ovulation.

Hypospadias A condition in which there is an abnormal, lengthened slit on the underside of the penis instead of the normal opening at the end.

Hypothalamus A small gland located at the base of the brain that secretes hormones that regulate various bodily functions, including ovulation in women and sperm production in men.

Hypothyroidism A metabolic disorder in which the thyroid gland produces lower-than-normal amounts of thyroid hormone.

Hysterosalpingography A procedure in which a special dye is introduced into a woman's cervix, uterus, and fallopian tubes, allowing examination of their size and shape.

Hysteroscopy A procedure in which the inside of the uterus is viewed through a long, slender, telescopelike instrument (the hysteroscope) inserted through the vagina and cervix.

Implantation bleeding Normal light bleeding or spotting that sometimes occurs when a fertilized egg implants in the uterus.

Impotence A condition in which a man cannot achieve or sustain an erection long enough to ejaculate inside a woman's vagina.

Incompetent cervix A condition in which a pregnant woman's cervix begins to dilate too soon, causing a miscarriage.

Indemnity See *fee-for-service.*

Individual practice association (IPA) A type of managed care health insurance plan in which the insured person chooses a primary care physician from a preapproved list of providers and is referred for specialized services by the primary care physician as needed.

Infertility The absence of conception after at least one year of regular, unprotected intercourse.

Inflammatory bowel disease (IBD) A disorder of the intestines marked by recurrent episodes of diarrhea, fever, and abdominal pain, the causes of which are largely unknown.

Intended father The man who raises and cares for a child born through a surrogacy arrangement (*see biological father*).

Intended mother The woman who raises and cares for a child born through a surrogacy arrangement (*see biological mother* and *gestational mother*).

Intracytoplasmic sperm injection (ICSI) A technique in which a single sperm cell is injected through a microsurgical needle directly into the cytoplasm of an egg to facilitate fertilization.

Intrauterine growth retardation (IUGR) A condition in which a newborn is smaller than the size that is normal for the amount of time spent in the uterus.

Intrauterine insemination (IUI) A technique in which sperm are introduced directly into a woman's cervix or uterus to produce pregnancy, with or without ovarian stimulation to produce multiple ova.

In vitro fertilization (IVF) A form of assisted reproduction in which an egg and sperm are combined in a laboratory dish and the resulting embryo or preembryo is transferred into a woman's fallopian tube.

Laparoscopy A procedure in which a long, slender, telescopelike instrument is inserted through a small incision in a woman's abdomen in order to view the internal organs and structures.

Liquefaction The process by which normal semen thickens after ejaculation and then, after about thirty to forty minutes, returns to a liquid state.

Luteal phase The second half of the menstrual cycle, beginning at ovulation (day 14 in an average twenty-eight-day cycle) and ending with menstruation (day one of the next menstrual cycle).

Luteinizing hormone (LH) A hormone that is secreted by the pituitary and that, with follicle-stimulating hormone (FSH), signals the testes to produce sperm in males and the ovaries to produce ova in females.

Managed care An arrangement for providing health care in which the insured person uses preapproved care providers and pays a set monthly premium in addition to a small fee (usually five to fifteen dollars) at the time services are rendered.

Medroxyprogesterone A synthetic form of the hormone progesterone, used to trigger menses.

Metformin An anti-inflammatory drug sometimes used with ovulation-inducing drugs.

Microinsemination A procedure in which sperm are concentrated into a small drop of fluid and placed around the eggs to increase the chances of fertilization.

Microsurgical epididymal sperm aspiration (MESA) A procedure in which actively moving sperm are collected directly from the epididymis, the organ next to each testis where sperm mature and are stored, usually for use in an IVF procedure.

Miscarriage Spontaneous loss of a pregnancy before twenty weeks of gestation.

Morphology In a semen analysis, the physical appearance of sperm cells.

Morula The zygote at about two to three days after fertilization, when it has divided into sixteen cells.

Motility In a semen analysis, the degree to which sperm cells are able to spontaneously propel themselves.

Multifetal pregnancy reduction See selective reduction.

Neonatal intensive care unit (NICU) A specialized nursery for newborns who require close medical attention and monitoring immediately after birth.

Oligospermia A cause of male infertility in which fewer than four million sperm are present in the semen from one ejaculation.

Oocyte An immature ovum; the egg before it is released at ovulation.

Ova [sing., *ovum*] The female sex cells, or eggs, which are produced in the ovaries.

Ovarian hyperstimulation syndrome (OHSS) A rare side effect that may occur when gonadotropins are used for ovulation induction, in which the ovaries become enlarged and tender.

Ovaries [sing., *ovary*] Two small organs on either side of a woman's lower pelvis which produce *ova*, or eggs, and hormones.

Ovulation induction A procedure in which medication is used to stimulate a woman's ovaries to produce multiple mature follicles and ova.

Pelvic inflammatory disease (PID) Inflammation of the female upper reproductive tract (uterus, tubes, and ovaries) usually resulting from infection with chlamydia and/or gonorrhea.

Penis The male reproductive organ, through which semen exits during ejaculation.

Perinatologist A physician specializing in the care of pregnant women and their babies during labor and delivery.

Peritoneum The space inside the abdomen holding the body's internal organs.

Pituitary A gland located at the base of the brain that is triggered by hormones released from the hypothalamus to secrete hormones that regulate various bodily functions, including ovulation in women and sperm production in men.

Pituitary adenoma An abnormal growth on the pituitary gland, sometimes indicated by high prolactin levels.

Placenta The thick pad of tissue inside a pregnant woman's uterus that provides nourishment to and disposes of waste from the growing fetus.

Polycystic ovarian syndrome (PCO) A condition in which multiple cysts form on one or both ovaries, preventing ovulation.

Polyps See *uterine polyps*.

Postcoital test (PCT) A test used to evaluate the interaction between a man's sperm and a woman's cervical mucus.

Postpartum hemorrhage (PPH) Excessive blood loss immediately after delivery.

Preeclampsia High blood pressure during pregnancy.

Preferred provider organization (PPO) A type of managed care health plan in which the insured person receives health care through a network of physicians and hospitals that give the PPO discounts on their usual rates.

Pregnancy-induced hypertension See *preeclampsia*.

Premature ejaculation A condition in which ejaculation occurs before the penis enters a woman's vagina.

Premature ovarian failure A condition in which a woman's body stops producing estrogen and ceases ovulating before the age of 40 (also called *early menopause*).

Preterm birth, preterm delivery Delivery of a baby before the end of the thirty-seventh week of gestation.

Preterm labor Labor that begins before the end of the thirty-seventh week of gestation.

Primary infertility Infertility in a couple who has never had a pregnancy.

Progesterone A hormone that is produced in a woman's ovaries and that stimulates the endometrium to thicken in preparation for possible pregnancy during the latter part of the menstrual cycle.

Prolactin A hormone that is secreted by the pituitary and stimulates milk production during pregnancy.

Pronuclear stage transfer (PROST) *See zygote intrafallopian transfer.*

Pronucleus The fertilized ovum before it has begun the process of cell division.

Prostaglandins Hormones that cause the muscles of a woman's uterus to contract, producing cramps.

Prostate gland A gland that is located just below a man's bladder and secretes fluid that helps sperm pass through the urethra.

Prostatitis Infection of the prostate gland.

Reproductive endocrinologist A physician who specializes in diagnosing and treating infertility.

Respiratory distress syndrome (RDS) A condition in which a newborn has breathing problems because the lungs are not fully mature at birth.

Retrograde ejaculation A condition in which semen backs up into the bladder instead of exiting out through the urethra.

Round spermatid nuclear injection (ROSNI) A form of ART in which the nucleus of an immature sperm cell (a round spermatid) is isolated and injected into an ovum.

Scrotum A saclike pouch containing the testes at the base of the penis.

Secondary infertility Infertility in a couple who has had one or more pregnancies.

Selective reduction, fetal reduction, selective birth A procedure in which one or more fetuses is intentionally removed during a multiple pregnancy in order to increase the chances of survival of the remaining fetus(es).

Seminal vesicle A small gland that is located just behind the bladder in the male and secretes fluid that helps to lubricate sperm.

Serum The liquid portion of blood that remains after the solid components have been removed.

Sexually transmitted disease (STD) An infection that is spread by sexual contact.

Sickle cell anemia An inherited disorder in which the red blood cells are abnormally shaped and therefore cannot function properly; the disease seems to be associated with decreased sperm production.

Singleton pregnancy A pregnancy in which one fetus develops inside the uterus.

Special needs adoption, special needs child Adoption of a child with physical and/or emotional problems, often arranged through an agency specializing in such placements.

Sperm cells The male sex cells, which are produced in the testes.

Sperm count An assessment of the number of sperm present in each milliliter of semen.

Sperm penetration assay (SPA) A test examining the ability of sperm to penetrate and fertilize an egg.

Sperm washing A procedure used to remove components other than sperm from a semen sample to be used for intrauterine insemination.

Surrogacy An arrangement in which a woman (the surrogate) carries a pregnancy for an infertile person or couple.

Testes Two small organs that are located at the base of the male's penis and in which sperm are produced.

Testicles See testes.

Testicular sperm extraction (TESE) A procedure in which immature sperm cells are collected directly from the testes through a fine needle.

Testosterone A hormone that is produced by a man's testes and that helps to maintain the production of sperm.

Therapeutic donor insemination (TDI) A type of intrauterine insemination in which sperm from a donor, either known or anonymous, is used.

Third-party reproduction The use of sperm, ova, embryos, or the uterus of a third person by a couple wishing to have a child.

Thyroid A gland located at the base of the neck which secretes hormones influencing growth and metabolism.

Thyroid-stimulating hormone A hormone that is secreted by the pituitary and stimulates the growth and function of the thyroid gland.

Tubal embryo transfer (TET) A variation of IVF in which one or more fertilized eggs are transferred into a woman's fallopian tube at the four- to eight-cell stage of development.

Tubal ligation A type of female sterilization in which the fallopian tubes are cut, clipped, or tied in order to prevent pregnancy.

Tubal ligation reversal (TLR) A microsurgical procedure in which a woman who has previously had surgical sterilization has one or both fallopian tubes reconnected to allow ovulation and pregnancy.

Ultrasound A procedure in which sound waves are used to create an image of the internal structures and organs.

Undescended testicle See cryptorchidism.

Unexplained infertility Infertility for which the cause cannot be determined with currently available diagnostic techniques.

Urethra A narrow, tubelike structure through which urine passes on its way from the bladder to the outside of the body; in males, it is also a passageway for sperm.

Uterine fibroids Abnormal, benign (noncancerous) growths attached either inside or outside the wall of a woman's uterus.

Uterine polyps Abnormal, benign (noncancerous) growths attached to a short, stalklike structure that protrudes from the outer or inner surface of a woman's uterus.

Uterine septum A thin wall of tissue inside the uterus that divides the uterine cavity into two spaces and that usually can be removed surgically.

Uterus The hollow, muscular organ in a woman's lower abdomen, in which a developing fetus grows during pregnancy.

Vagina The elastic, muscular passageway leading from the cervix to the outside of a woman's body.

Vaginismus A condition in which the muscles in and around the vagina undergo painful spasms when vaginal penetration is attempted.

Varicocele A cause of male infertility in which varicose veins are present in the blood vessels above the testes.

Vas deferens The long, narrow tube through which sperm pass on their way out of the body during ejaculation.

Vasectomy A procedure used for male sterilization, in which a small segment of each ductus deferens is surgically removed to prevent sperm from entering the ejaculate.

Vertex presentation A situation in which a baby is in the normal, head-down position at the time of labor and delivery.

Volume measurement In a semen analysis, a measurement of how many milliliters of semen are collected in the ejaculate.

Zona-free hamster ova penetration test A test in which the ability of a man's sperm to fertilize a human ovum is determined by combining the sperm with hamster ova from which the outer membrane (zona pellucida) has been removed.

Zona pellucida The outer membrane of an ovum, which must be penetrated by a sperm cell for fertilization to take place.

Zygote See pronucleus.

Zygote intrafallopian transfer (ZIFT) A variation of IVF in which one or more eggs are fertilized in a laboratory dish and transferred into a woman's fallopian tube before cell division has begun. This procedure is sometimes called *pronuclear stage transfer (PROST)*.

Index

Marion Powell Women's Health Information Centre
Women's College Ambulatory Care Centre
76 Grenville Street, Room 916
Toronto ON M5S 1B2
416-323-6045

Women's Health Information Centre
Women's College Ambulatory Care Centre
76 Grenville Street, Room 916
Toronto ON M5S 1B2
416-323-6045

07/18/05